Lecture Notes in Mathematics

Edited by A. Dold and B. Eckmann

738

P. E. Conner

Differentiable Periodic Maps

Second Edition

Springer-Verlag
Berlin Heidelberg New York 1979

Author

P. E. Conner
Mathematics Department
Louisiana State University
Baton Rouge, Louisiana 70803
U.S.A.

AMS Subject Classifications (1970): 57S20, 55N22, 57R75

ISBN 3-540-09535-7 Springer-Verlag Berlin Heidelberg New York
ISBN 0-387-09535-7 Springer-Verlag New York Heidelberg Berlin

First edition published as P. E. Conner/E. E. Floyd, Differentiable Periodic Maps (Ergebnisse der Mathematik und ihrer Grenzgebiete, Band 33) Berlin-Göttingen-Heidelberg: Springer 1964.
ISBN 3-540-03125-1 Springer-Verlag Berlin Heidelberg New York

Library of Congress Cataloging in Publication Data
Conner, Pierre E
Differentiable periodic maps.
(Lecture notes in mathematics; 738)
Bibliography: p.
Includes index.
1. Topological transfomation groups. 2. Cobordism theory. 3. Differentiable mappings.
I. Title. II. Series: Lecture notes in mathematics (Berlin); 738.
QA3.L28 no. 738 [QA613.7] 510'.8s [514'.7] 79-19135
ISBN 0-387-09535-7

Printing and binding: Beltz Offsetdruck, Hemsbach/Bergstr.
2141/3140-543210

CONTENTS

INTRODUCTION

For the second edition of this book we have adopted the viewpoint that it is to serve to introduce students to the techniques by which bordism can be applied to the study of periodic diffeomorphisms on closed manifolds. Other tools are available in the field of periodic maps. A good exposition of the classical P.A. Smith Theory can be found in Floyd's article in [B_4]. For an introduction to Equivariant K-Theory we may cite [ASe_2, Se] while the reference for the Atiyah-Singer-Segal G-Signature Theorem is [ASe_1]. Also equivariant surgery, guided by Browder, Quinn and Petrie, is developing into a most significant approach to periodic diffeomorphisms. Of course all the paths cross frequently; for example, tom Dieck [D_1 - D_5] weaves together a combination of equivariant cobordism and equivariant K-theory. The reader will agree that Borel's suggestion [B_4] of the need of methods in transformation groups which use differentiability in a key fashion has been carried out.

In Chapter I we discuss, with virtually no change from the first edition [CF_2], except for notation, the oriented bordism functor $\{MSO_*(X,A),\varphi_*,\partial\}$ and the unoriented bordism functor $\{MO_*(X,A),\varphi_*,\partial\}$. Our motivation is the eventual application to periodic diffeomorphisms, however bordism has found wider applications in topology and for a study of bordism as a phenomenon in its own right we could refer the reader to [CS].

In Chapter II we take up involutions on unoriented manifolds. Most of the questions about this topic raised in the first edition have been answered and we have made an effort to include these new results. We point out specifically the work of Boardman [Ba] which is a major objective of this chapter. For another approach we point out the new paper [KSt]. We have kept the study of $C_2 \times C_2$ actions with isolated fixed points, however those results have been generalized by tom Dieck [D_4].

For Chapter III we concentrate on maps of odd prime power period on closed oriented manifolds. This chapter was rewritten in the light of [CF_3]. A tedious computation of $MSO_*(C_{p}r)$ from the first edition is omitted, although the structure of $MSO_*(C_p)$ is studied in detail. An introduction to some of the work of tom Dieck [D_5] confined here to the cyclic odd prime power case, has been added. Some preliminary material, now made quite obsolete by the work of Floyd and tom Dieck [F,D_2], on abelian p-group actions which appeared in the first edition has been deleted.

We recommend Stong's book [St_2] be used in conjunction with these notes It contains the requisite background material on differential topology and gives a thorough treatment of Thom's bordism concept. Also Bredon's book [Br_2] is an introduction to compact transformation groups. It is not possible to include all the results which have evolved in the fifteen years since the appearance of the first edition. It is not our intention to produce a new book; only a second edition in which the material, we feel, will serve the student as a good introduction.

Professor Floyd, now Dean of Faculty of the University of Virginia, kindly agreed to let us produce this second edition. We express our thanks for his friendship and for the long, and we believe fruitful, collaboration of Conner and Floyd. We express our grateful remembrance of the late Professor G.T. Whyburn who for so many years at Virginia supported our work. In the past we have benefitted immeasurably from the help extended to our research program by the Sloan Foundation and the National Science Foundation. The present manuscript was prepared during a sabbatical leave from LSU that was spent at the Mathematical Institute of the University of Heidelberg. As a final personal note we say to Deane Montgomery: "Thank your for everything."

P.E. Connèr
June, 1978

CHAPTER I. BORDISM GROUPS

Our fundamental objective in this preliminary chapter is the introduction and development in an elementary and indeed perhaps naive fashion of bordism as a generalized homology theory. We wish to convey the feeling of working with bordism so that we may apply it to periodic maps. Given a topological pair (X,A) we consider all maps $f:(B^n, \dot{B}^n) \to (X,A)$ where B^n is a smooth compact oriented n-dimensional manifold with boundary $\dot{B}^n$. We introduce a relation of bordism type on all such f so that in Section 4 we shall arrive at $MSO_n(X,A)$, the resulting set of equivalence classes, together with a natural abelian group structure. These groups are the oriented bordism groups of (X,A) and in Section 5 we shall find that these yield a generalized homology theory; that is, with the exception of the dimension axiom, the Eilenberg-Steenrod axioms for a homology theory are satisfied by oriented bordism. In Section 12 we interpret $MSO_n(X,A)$ as the homology theory of the pairs (X,A) with coefficients in the MSO-spectrum. Specifically, for a finite CW-pair there is an isomorphism of $MSO_n(X,A)$ with the homotopy group $\pi_{n+k}(X/A \wedge MSO(k))$ where MSO(k) is the Thom space of the universal oriented k-plane bundle and $k \geq n + 2$. In Section 13 we sketch the dual generalized cohomology theory, called oriented cobordism, due to Atiyah. Simply to establish conventional and notational rapport with the reader we shall mention some background material from differential topology.

1. Differentiable manifolds

In this section we outline properties of differentiable manifolds. We follow the custom of using *differentiable* and *smooth* as synonyms and both as an abbreviation for *differentiable of class* C^∞.

A map $f : A \to R^n$, with A a subset of Euclidean m-space, R^m, is said to be differentiable on A if and only if at each $x \in A$ there is an open neighborhood V of x in R^m and a smooth extension $F : V \to R^m$ of $f|V \cap A \to R^m$.

Let H^n denote the half-space in R^n consisting of all points $(x_1, x_2, \ldots, x_n)$ with $x_1 \geq o$. A separable metric space B^n is a topological n-manifold if and only if for each $x \in B^n$ there is an open neighborhood V of x which is homeomorphic to an open subset of H^n. A *differential structure* on a topological n-manifold is a collection of ordered pairs (V_i, h_i), where V_i is open in B^n and h_i is a homeomorphism of V_i onto an open subset of H^n such that

i) the collection $\{V_i\}$ forms an open covering of B^n

ii) for every pair (i,j) the map

$$h_j h_i^{-1} : h_i(V_i \cap V_j) \to R^n$$

is differentiable

iii) the collection $\{(V_i, h_i)\}$ is maximal with respect to the preceding two properties.

In practice it is customary to give a collection of pairs satisfying i) and ii) and then appeal to the fact that it is contained in a unique maximal collection. A topological n-manifold together with a differentiable structure is called a *differentiable n-manifold*. Without significant exception we shall be concerned with smooth compact manifolds.

The pairs (V_i, h_i) are referred to as *coordinate neighborhoods*. In the above definitions, H^o is taken to be a single point so that a compact 0-manifold is a finite set.

Let $R^{n-1} \subset H^n$, $n > o$, be the set of all $(x_1, \ldots, x_n)$ with $x_1 = o$. The *boundary*, $\dot{B}^n$, of a differentiable manifold then consists of those points $x \in B^n$ for which there is a coordinate pair (V,h) with $x \in V$ and $h(x) \in R^{n-1} \subset H^n$. If $\dot{B}^n$ is empty, then B^n is a manifold without boundary. In that case H^n can be replaced by R^n in the definition of differentiable structure. A compact differentiable manifold without boundary will be termed a *closed manifold*. The boundary, $\dot{B}^n$, of a differentiable n-manifold is always a differentiable (n-1)-manifold without boundary. Whenever (V,h) is a coordinate neighborhood of B^n then $(V \cap \dot{B}^n, h|V \cap \dot{B}^n)$ is to be a coordinate neighborhood of $\dot{B}^n$.

If U is an open subset of a differentiable n-manifold B^n then U naturally receives a differentiable structure; namely all the coordinate neighborhoods (V,h) of B^n with $V \subset U$. We call this the differentiable structure induced by that of B^n. The following easily verified remark is useful in piecing together structures.

(1.1) LEMMA: *Suppose that* U_1 *and* U_2 *are open neighborhoods of the topological n-manifold* B^n *for which* $U_1 \cup U_2 = B^n$. *If* U_1 *and* U_2 *both have differential structures which induce the same differentiable structure on* $U_1 \cap U_2$ *then there exists a unique differentiable structure on* B^n *which induces the differentiable structures of* U_1 *and* U_2. ∎

If M^n is a differentiable manifold without boundary and B^m a differentiable manifold, then the reader will see that there is naturally induced a product differential structure on $M^n \times B^m$. Even more generally, $B_1^n \times B_2^m \smallsetminus \dot{B}_1^n \times \dot{B}_2^m$ has a natural differentiable structure.

A map $\varphi : B_1^m \to B_2^n$ between differentiable manifolds is *differentiable* if and only if whenever (V,h) and (W,k) are coordinate neighborhoods of B_1^m and B_2^n respectively then $k\varphi h^{-1} : h(V \cap \varphi^{-1}(W)) \to R^n$ is differentiable. The manifolds B_1^n and B_2^n are *diffeomorphic* if and only if there is a homeomorphism f of B_1^n onto B_2^n with both f and f^{-1} differentiable.

We shall assume the *differentiable collaring theorem*, $[M_6, St_2]$.

(1.2) THEOREM: *For a differentiable manifold* B^n *there is an open set* U *containing* $\dot{B}^n$ *and a diffeomorphism* φ *of* U *onto* $\dot{B}^n \times [o,1)$ *with* $\varphi(x) = (x,o)$ *for* $x \in \dot{B}^n$. ∎

We shall close this section with a brief discussion of *orientability* for compact manifolds. If B^n is compact and connected then the singular homology group $H_n(B^n, \dot{B}^n; Z)$ is either Z or it is trivial. In the former case B^n is said to be orientable. This is equivalent to the reducibility of the structure group of the tangent bundle of B^n from $O(n)$ to the subgroup $SO(n)$. An *orientation* on B^n is the choice of a generator $\sigma \in H_n(B^n, \dot{B}^n; Z) \simeq Z$. Then σ is referred to as the *fundamental homology class* of the oriented manifold. If B^n is not connected then it is said to be orientable if and only if each component is so. To orient such a manifold is to choose a generator $\sigma_c \in H_n(B_c^n, \dot{B}_c^n; Z)$ for each component of $H_n(B^n, \dot{B}^n; Z)$ which is $\sum_c H_n(B_c^n, \dot{B}_c^n; Z)$. Now the *orientation class* (fundamental homology class) $\sigma(B^n)$ of the oriented manifold is the element of $H_n(B^n, \dot{B}^n; Z)$ given by $\sum \sigma_c$. An orientation of B^n induces an orientation of $\dot{B}^n$ by assigning $\partial\sigma(B^n)$, where $\partial : H_n(B^n, \dot{B}^n; Z) \to H_{n-1}(\dot{B}^n; Z)$ is the boundary homomorphism in the exact sequence of the pair. A diffeomorphism $\varphi : B_1^n \to B_2^n$ between oriented manifolds is *orientation preserving* if and only if $\varphi_* \sigma(B_1^n) = \sigma(B_2^n)$.

2. The Thom bordism groups

In this section we define and give elementary properties of the Thom groups MSO_n and MO_n. At the end of the section we summarize briefly the deeper structure of these groups, but it will be some time before that information is actually used. Hereafter in this book we shall use *manifold* for *differentiable manifold*.

Given a closed oriented manifold, M^n, then $-M^n$ is the oriented manifold obtained by using $-\sigma(M^n)$ as orientation class. Given closed oriented manifolds M_1^n and M_2^n we call the *disjoint union* of M_1^n and M_2^n any compact oriented manifold which is a disjoint union $M_1'^n \sqcup M_2'^n$ of closed oriented submanifolds with $M_i'^n$ diffeomorphic to M_i^n via an orientation preserving diffeomorphism. Denote a disjoint union simply by $M_1^n \sqcup M_2^n$.

A closed oriented manifold M^n is said to *bord* if and only if there is a compact oriented manifold B^{n+1} for which $\dot{B}^{n+1}$ is diffeomorphic to M^n via an orientation preserving diffeomorphism. Two closed oriented manifolds M_1^n and M_2^n are *bordant* if and only if the disjoint union $M_1^n \sqcup -M_2^n$ bords; we refer to this as the *bordism relation*.

(2.1) THEOREM: *The bordism relation is an equivalence relation on the oriented diffeomorphism classes of closed oriented* n-*manifolds. The resulting set* MSO_n *of equivalence classes is an abelian group with addition induced by disjoint union.*

PROOF: To see that M^n is bordant to itself, form the oriented product manifold $I \times M^n$ where I is the oriented unit interval. Then $(I \times M^n)^{\bullet} = \dot{I} \times M^n = \{1\} \times M^n \sqcup \{o\} \times -M^n$. Hence the disjoint union $M^n \sqcup -M^n$ bords. The bordism relation is clearly symmetric. It is in the proof of transitivity that the collaring theorem (1.2) plays its typical role. Suppose that M_1^n is bordant to M_2^n and that M_2^n is bordant to M_3^n. Then there are compact oriented manifolds B_1^{n+1} and B_2^{n+1} with $\dot{B}_1^{n+1} = M_1^n \sqcup -M_2^n$ and $\dot{B}_2^{n+1} = M_2^n \sqcup -M_3^n$. We may suppose $B_1^{n+1} \cap B_2^{n+1} = M_2^n$. Now set $B^{n+1} = B_1^{n+1} \cup B_2^{n+1}$; then B^{n+1} is a topological oriented (n+1)-manifold with $\dot{B}^{n+1} = M_1^n \sqcup -M_3^n$. It is necessary to give B^{n+1} a differentiable structure. By (1.2) there is an open set U_1, of B^{n+1}, containing M_2^n and a homeomorphism h of U_1 onto $M_2^n \times (-1,1)$ with $h|U_1 \cap B_1^{n+1} \to M_2^n \times (-1,o]$ a diffeomorphism and $h|U_1 \cap B_2^{n+1} \to M_2^n \times [o,1)$ also a diffeomorphism. Via h we give to U_1 the differential structure from $M_2^n \times (-1,1)$. Next let $U_2 = B^{n+1} \setminus M_2^n$ so that U_2 has a natural differential structure.

By (1.1) we now receive a differentiable structure on B^{n+1} and the bordism relation is seen to be transitive.

The equivalence class *(oriented bordism class)* to which M^n belongs is denoted by $[M^n]$ and the collection of all such classes by MSO_n. An abelian group structure is put on MSO_n by disjoint union; that is, $[M_1^n] + [M_2^n] = [M_1^n \sqcup M_2^n]$. The additive identity is the class of those manifolds which do bord; moreover, $-[M^n] = [-M^n]$. ∎

The direct sum $MSO_* = \sum MSO_n$ can be given the structure of a graded commutative ring with identity. The product of the homogeneous elements $[M_1^n]$ and $[M_2^m]$ is given by $[M_1^n][M_2^m] = [M_1^n \times M_2^m]$. This is seen to be a well defined associative and distributive product. Furthermore $[M_1^n][M_2^m] = (-1)^{nm}[M_2^m][M_1^n]$.

The groups MSO_n have been completely determined. Thom showed in his original work on this subject $[T_2]$ that $MSO_* \otimes Q$, Q the rationals, is a polynomial algebra over Q and the generators may be taken to be the bordism classes of the complex projective spaces $CP(2k)$, $k=1,2,\ldots$. Milnor $[M_5]$, and independently Averbuch [Av], showed that MSO_* has no odd torsion (verifying Thom's conjecture). The 2-primary torsion was then completely determined by Wall [Wa], who showed every non-trivial torsion class has order 2. In so doing he found it useful to work with the elements of order two discovered earlier by Dold, $[Do_2]$. In the meantime, Milnor $[M_5]$ had also given the multiplicative structure of MSO_*/Tor, Tor here is the torsion ideal. Namely, there exist closed oriented manifolds $[Y^{4k}]$, $k = 1,2,\ldots$, such that MSO_*/Tor is the graded polynomial ring over Z generated by $[Y^{4k}]$. If $2k+1$ is an odd prime, p, then one may choose $Y^{2p-2} = CP(p-1)$, complex projective $(p-1)$-space.

In addition to oriented bordism there is also an *unoriented bordism theory*. In the unoriented theory all closed manifolds are used; no orientability considerations are imposed. The unoriented bordism class of M^n is denoted by $[M^n]_2$ and the set of unoriented bordism classes is denoted by MO_n. As in MSO_n this MO_n has an abelian group structure defined by disjoint union, however every element in MO_n has order 2. The direct sum $MO_* = \sum MO_n$ is, via cartesian products, a graded commutative algebra over $Z/2Z$. Thom $[T_2]$ showed that MO_* is a graded polynomial algebra over $Z/2Z$ with a generator in each dimension n which is not of the form $2^j - 1$. In even dimensions the generator can be chosen to be the unoriented bordism class of the real projective space $RP(2m)$; an explici list of generators for the odd dimensions was first given by Dold $[Do_2]$.

Beginning with Thom, the approach to the computation of bordism groups had, until recently, rested upon reinterpreting the groups as the homotopy groups of the corresponding Thom spectrum and then from the study of the module structure of the cohomology of the spectrum over the Steenrod algebra an analysis of the homotopy groups was made. There is a full exposition of this approach in Stong [St_2]. More recently, Quillen has given a new approach based on the use of formal group laws. This is especially successful in determining the structure of the weakly complex bordism ring MU_*,[Q_1, Q_2].

3. Straightening angles

This is a little digression, following Milnor's exposition [M_9,St_2] to show the cartesian product of two differentiable manifolds can itself be given a differentiable structure even though both of the two factors have boundaries. More or less by continuation, this section will be understood to apply to any of those places in the book where "rounding" or "smoothing" a corner is required.

Let $R_+ \subset R$ consist of all non-negative real numbers. Pick once and for all a homeomorphism τ of the quadrant $R_+ \times R_+$ onto the half-space $R \times R_+$ with τ a diffeomorphism of $R_+ \times R_+ \smallsetminus (o,o)$ onto $R \times R_+ \smallsetminus (o,o)$. For example, in polar coordinates let $\tau(r,\Theta) = (r,\Theta) = (r,2\Theta)$ for $o \leq \Theta \leq \pi/2$.

Suppose now we have a topological manifold B^n and that also we have in B^n a submanifold M^{n-2} without boundary, closed in B^n, such that

1) $B^n \smallsetminus M^{n-2}$ has a differentiable structure

2) M^{n-2} has a differentiable structure

3) there is an open neighborhood U of M^{n-2} in B^n and a homeomorphism Φ of U onto $M^{n-2} \times R_+ \times R_+$ with $\Phi(x) = (x,o,o)$ for all $x \in M^{n-2}$ and with Φ a diffeomorphism of $U \smallsetminus M^{n-2}$ onto $M^{n-2} \times R_+ \times R_+ \smallsetminus M^{n-2} \times o \times o$.

Let $\tau' : M^{n-2} \times R_+ \times R_+ \to M^{n-2} \times R \times R_+$ be given by $\tau'(x,y,z) = (x,\tau(y,z))$. We then have a composite homeomorphism:

$$\tau'\Phi : U \to M^{n-2} \times R \times R_+.$$

There is a product differentiable structure on U such that $\tau'\Phi$ is a diffeomorphism. Hence U and $B^n \smallsetminus M^{n-2}$ have differentiable structures and these are seen to induce the same differentiable structure on their intersection $U \cap (B^n \setminus M^{n-2})$. Use of (1.1) now imparts a differentiable structure to B^n. This process is referred to as *straightening the angle*.

Consider for example differentiable manifolds B_1^m and B_2^n. As pointed out in Section 1, $B_1^m \times B_2^n \setminus (\dot{B}_1^m \times \dot{B}_2^n)$ has a natural differentiable structure By the collaring theorem (1.2), $\dot{B}_1^m$ and $\dot{B}_2^n$ have neighborhoods U_1 and U_2 in B_1^m and B_2^n respectively with diffeomorphisms Φ_1 of U_1 with $\dot{B}_1^m \times R_+$ and Φ_2 of U_2 with $\dot{B}_2^n \times R_+$. Let $U = U_1 \times U_2$; then $\Phi = \Phi_1 \times \Phi_2$ is a homeomorphism of U onto $\dot{B}_1^m \times \dot{B}_2^n \times R_+ \times R_+$ with the properties of 3 above. We then obtain a differentiable structure on the product $B_1^m \times B_2^n$ by straightening the angle.

The next lemma will be used in Section 5.

(3.1) LEMMA: *Suppose* P *and* Q *are closed disjoint subsets of the compact n-manifold* B^n. *There exists a compact topological submanifold* $B_1^n \subset B^n$ *with* $P \subset B_1^n$, $Q \cap B_1^n = \emptyset$ *and* B_1^n *closed in* B^n. *Moreover* B_1^n *can be given a differentiable structure by straightening the angle.*

PROOF: By (1.2) we may as well identify a neighborhood U of $\dot{B}^n$ with $\dot{B}^n \times [0,1)$. By normality there exist disjoint closed subsets P_1 and Q_1 in $\dot{B}^n$ containing $P \cap \dot{B}^n$ and $Q \cap \dot{B}^n$ in their respective interiors. By compactness there is a t, $o < t < 1$, such that

$$P_1 \times [o,t) \supset P \cap (\dot{B}^n \times [o,t)).$$

$$Q_1 \times [o,t) \supset Q \cap (\dot{B}^n \times [o,t)).$$

Let W^n be the n-manifold $B^n \setminus (\dot{B}^n \times [o,t))$, and let $P' = P \cap W^n$, $Q' = Q \cap W^n$. There exists a smooth function $f' : W^n \to [0,1]$ with

$$f'((P_1 \times t) \cup P') = o$$

$$f'((Q_1 \times t) \cup Q') = 1.$$

Extend f' to a function $f : B^n \to [o,1]$ defining $f(x,s) = f'(x,t)$ for $(x,s) \in \dot{B}^n \times [o,1)$. Then $f(P) = o$, $f(Q) = 1$ and f is differentiable on $\dot{B}^n \times [o,t]$. There exists a differentiable approximation F to f with $F = f$ on $\dot{B}^n \times [o,t_o]$ where $o < t_o < t$. We may suppose F chosen so close to f that g.l.b. $F(Q)$ > l.u.b. $F(P)$. We can then find s_o with l.u.b. $F(P) < s_o <$ g.l.b. $F(Q)$ and so that s_o is a regular value of F [Mr]. Now $F^{-1}[o,s_o]$ is seen to be a topological n-manifold B_1^n with $P \subset B_1^n$ and $B_1^n \cap Q = \emptyset$. It is not difficult to see that the angle can be straightened around the (n-2)-manifold $F^{-1}(s_o) \cap \dot{B}^n$ to yield a differential structure on B_1^n. ∎

4. The oriented bordism functor

In this section we define a singular bordism theory analogous to, but more naive than, singular homology. Fix a topological pair (X,A), then a *singular manifold* in (X,A) is a pair (B^n,f) consisting of a compact oriented manifold B^n and a map $f : (B^n,\dot{B}^n) \to (X,A)$. If $A = \emptyset$ then of course $\dot{B}^n = \emptyset$. An oriented singular manifold (B^n,f) in (X,A) is said to bord if and only if there is a compact oriented manifold C^{n+1} and a map $F : C^{n+1} \to X$ for which

i) B^n is contained in $\dot{C}^{n+1}$ as a regular submanifold whose orientation is induced by that of C^{n+1};

ii) $F|B^n = f$ and $F(\dot{C}^{n+1} \setminus B^n) \subset A$.

From two oriented singular manifolds (B^n_1,f_1) and (B^n_2,f_2) a disjoint union is defined, $(B^n_1 \sqcup B^n_2,\ f_1 \sqcup f_2)$ wherein $B^n_1 \cap B^n_2 = \emptyset$, $f_1 \sqcup f_2|B^n_1 = f_1$ and $f_1 \sqcup f_2|B^n_2 = f_2$. Define $-(B^n,f) = (-B^n,f)$, then we may say (B^n_1,f_1) and (B^n_2,f_2) are bordant if and only if the disjoint union $(B^n_1 \sqcup -B^n_2,$ $f_1 \sqcup f_2)$ bords in (X,A). We leave it to the reader to show this defines an equivalence relation among the oriented singular manifolds in (X,A); an argument similar to straightening angles is needed for transitivity.

Denote the oriented bordism class of (B^n,f) by $[B^n,f]$, and the collection of all such bordism classes by $MSO_n(X,A)$. An abelian group structure is imposed on $MSO_n(X,A)$ by disjoint union; that is,

$$[B^n_1,f_1] + [B^n_2,f_2] = [B^n_1 \sqcup B^n_2, f_1 \sqcup f_2].$$

The class of all (B^n,f) which do bord is the identity element, and of course $-[B^n,f] = [-B^n,f]$. We refer to $MSO_n(X,A)$ as the *n-dimensional oriented bordism group of* (X,A). Similar groups have been defined by Atiyah [A], Eels and Milnor. In one way or another the rest of the book is devoted to the study of these groups.

There is an MSO_*-module structure defined on the direct sum $MSO_*(X,A) = \sum MSO_n(X,A)$. From an oriented singular manifold (B^n,f) in (X,A) and a closed oriented manifold M^m a new singular manifold $(B^n \times M^m,g)$ may be defined by $g(x,y) = f(x)$. The module structure is now given by

$$[B^n,f][M^m] = [B^n \times M^m,g].$$

We leave to the reader the proof that this defines on $MSO_*(X,A)$ the structure of a graded right MSO_*-module.

Given a map

$$\varphi : (X,A) \to (X_1,A_1)$$

there is induced a natural homomorphism $\varphi_* : MSO_n(X,A) \to MSO_n(X_1,A_1)$ which is defined by $\varphi_*[B^n,f] = [B^n,\varphi f]$. There is also the homomorphism $\partial : MSO_n(X,A) \to MSO_{n-1}(A)$ which is given by $\partial[B^n,f] = [\dot{B}^n,f|\dot{B}^n]$. It is easily seen that ∂ is well defined and additive. Indeed $\varphi_* : MSO_*(X,A) \to MSO_*(X_1,A_1)$ and $\partial : MSO_*(X,A) \to MSO_*(A)$ are MSO_*-module homomorphisms of degree o and -1 respectively.

5. The Eilenberg-Steenrod axioms

In Section 4 we just introduced a covariant functor $\{MSO_n(X,A,\varphi_*,\partial\}$ on the category of topological pairs and maps. We wish to show it to be a generalized homology theory. This motion was studied formally by G.W. Whitehead [Wh] in 1962 and the concept of generalized homology is now routinely accepted as one of the principal tools of algebraic topology.

(5.1) THEOREM: *On the category of topological pairs and maps the oriented bordism functor* $\{MSO_n(X,A),\varphi_*,\partial\}$ *satisfies the first six Eilenberg-Steenrod axioms for a homology theory. However, for a single point we have* $MSO_n(pt) \simeq MSO_n$*, the oriented Thom bordism group.*

We proceed to enumerate the axioms; the first three are trivially verified.

(5.2) LEMMA: *If* $i : (X,A) \to (X,A)$ *is the identity map then* $i_* : MSO_n(X,A) \to MSO_n(X,A)$ *is the identity automorphism.* ∎

(5.3) LEMMA: *If* $\varphi : (X,A) \to (X_1,A_1)$ *and* $\Theta : (X_1,A_1) \to (X_2,A_2)$ *are maps then* $(\Theta\varphi)_* = \Theta_*\varphi_*$. ∎

(5.4) LEMMA: *For any map* $\varphi : (X,A) \to (X_1,A_1)$ *the diagram*

$$\begin{array}{ccc} MSO_n(X,A) & \xrightarrow{\partial} & MSO_{n-1}(A) \\ \downarrow{\scriptstyle \varphi_*} & & \downarrow{\scriptstyle (\varphi|A)_*} \\ MSO_n(X_1,A_1) & \xrightarrow{\partial} & MSO_{n-1}(A_1) \end{array}$$

is commutative. ∎

We have next to consider the homotopy axiom.

(5.5) LEMMA: *If* $\varphi_0, \varphi_1 : (X,A) \to (X_1,A_1)$ *are homotopic then* $\varphi_{0_*} = \varphi_{1_*}$.

PROOF: Let $h : (I \times X,\ I \times A) \to (X_1,A_1)$ be a homotopy between φ_0 and φ_1. For (B^n,f) a singular manifold in (X,A) define $\Theta : I \times B^n \to X_1$ by $\Theta(t,x) = h(t,f(x))$; then $\Theta(o,x) = \varphi_0 f(x)$ and $\Theta(1,x) = \varphi_1 f(x)$. Now from Section 3, $I \times B^n$ is a manifold and $(I \times B^n)^{\cdot} = (\dot{I} \times B^n) \cup (I \times -\dot{B}^n)$. Thus the disjoint union $1 \times B^n \sqcup o \times -B^n$ is a regular submanifold of the boundary and $\Theta(I \times -\dot{B}^n) \subset A_1$. Hence $[B^n,\varphi_0 f] = [B^n,\varphi_1 f]$ as required. ∎

One remark is needed before proving exactness. Let $V^n \subset M^n$ be a compact regular submanifold with boundary in a closed manifold M^n. If $f : M^n \to X$ is a map with $f(M^n \setminus \mathrm{Int}(V)) \subset A$, then $[M^n,f] = [V^n,f|V^n]$ in $MSO_n(X,A)$. This is seen as follows. Let $F : I \times M^n \to X$ be given by $F(t,x) = f(x)$. Now $(I \times M^n)^{\cdot} = \dot{I} \times M^n$ and $1 \times V^n \sqcup o \times -M^n$ is a regular submanifold of the boundary. Since $F(1 \times (M^n \setminus \mathrm{Int}(V))) \subset A$ we have $[M^n,f] = [V^n,f|V^n$ in $MSO_n(X,A)$. ∎

(5.6) LEMMA: *For every pair* (X,A) *the sequence*

$$\cdots \longrightarrow MSO_n(A) \xrightarrow{i_*} MSO_n(X) \xrightarrow{j_*} MSO_n(X,A) \xrightarrow{\partial} MSO_{n-1}(A) \longrightarrow \cdots$$

is exact.

PROOF: It is easy to verify that $\partial j_* = o$ and $i_*\partial = o$. Consider $[M^n,f] \in MSO_n(A)$. Apply the preceding remark with $V^n = \emptyset$ to see that $j_*i_*[M^n,f] = o$ in $MSO_n(X,A)$. Next consider $[V^n,f] \in MSO_n(X,A)$ which is also in the kernel of ∂. By definition then there is a compact oriented manifold B^n and a map $g : B^n \to A$ with $\dot{B}^n = \dot{V}^n$ and $g|\dot{B}^n = f|\dot{V}^n$. Identify V^n and $-B^n$ along their common boundary to obtain a closed oriented manifold M^n and a map $F : M^n \to X$ with $F|V^n = f$ and $F|B^n = g$. Now $[M^n,F] \in MSO_n(X)$ and the equality $j_*[M^n,F] = [V^n,F]$ in $MSO_n(X,A)$ follows from the remark preceding (5.6). The remainder of exactness is trivial. ∎

Finally we come to the (weak) excision axiom.

(5.7) LEMMA: *If* $\bar{U} \subset \mathrm{Int}(A)$, *then the inclusion* $i : (X \setminus U,\ A \setminus U) \subset (X,A)$ *induces an isomorphism*

$$i_* : MSO_n(X \setminus U,\ A \setminus U) \simeq MSO_n(X,A).$$

PROOF: We shall only show that i_* is an epimorphism; the remainder of the argument is similar. Let (B^n,f) be an oriented singular manifold in (X,A). Let $P = f^{-1}(X \setminus \mathrm{Int}(A))$ and $Q = f^{-1}(\bar{U})$. There exists by (3.1) a topological submanifold $B_1^n \subset B^n$ with $P \subset B_1^n$ and $B_1 \cap Q = \emptyset$. Furthermore B_1^n can be given a differentiable structure by straightening the angle. Now $[B_1^n, f|B_1^n]$ lies in $MSO_n(X \setminus U, A \setminus U)$ and just as in the remark preceding (5.6) it follows that $i_*[B_1^n, f|B_1^n] = [B^n, f]$ in $MSO_n(X,A)$. ■

6. Consequences of the axioms

A number of facts follow immediately from the first six Eilenberg-Steenrod axioms alone. Here the general reference would be to the book [ES]. For one thing we now have reduced bordism groups. We identify the Thom group MSO_n with the bordism group of a point $MSO_n(pt)$. The map, ε, which collapses X to a point then induces the augmentation homomorphism $\varepsilon_* : MSO_n(X) \to MSO_n(pt) = MSO_n$ and the reduced group $\widetilde{MSO}_n(X)$ is defined to be the kernel of ε_*. As usual $MSO_n(X) \simeq MSO_n \oplus \widetilde{MSO}_n(X)$. An oriented singular manifold (M^n,f) in X represents an element of the reduced group if and only if $[M^n] = o \in MSO_n$. We may observe too that for every triple $X \supset A \supset B$ there is an exact sequence [ES]

$$\ldots \longrightarrow MSO_n(A,B) \longrightarrow MSO_n(X,B) \longrightarrow MSO_n(X,A) \longrightarrow \ldots$$

Since oriented bordism is analogous to singular homology we should expect that by restricting ourselves to CW-pairs we would obtain improved statements about the bordism functor. Full excision will hold in this category and if $\varphi : (X,A) \to (X_1,A_1)$ is a relative homeomorphism between CW-pairs then $\varphi_* : MSO_n(X,A) \simeq MSO_n(X_1,A_1)$. In particular, $MSO_n(X,A) \simeq \widetilde{MSO}_n(X/A)$. When $A = \emptyset$ we shall agree that $X/\emptyset$ is the disjoint union of X and a point. There is also a Mayer-Vietoris sequence and a relative Mayer-Vietoris sequence for CW-triads.

A few groups may now be computed such as

$$\widetilde{MSO}_n(S^k) \simeq MSO_{n-k}$$

$$MSO_n(S^k) \simeq MSO_n \oplus MSO_{n-k}$$

and

$$MSO_n(I^k, S^{k-1}) \simeq \widetilde{MSO}_{n-1}(S^{k-1}) \simeq MSO_{n-k}.$$

These statements can also be interpreted along the following lines. If S^k is regarded as the oriented boundary of the (k+1)-disk then the identity map surely defines an element $\alpha = [S^k, id]$ in $\widetilde{MSO}_k(S^k)$. As a module over MSO_* the reduced bordism $\widetilde{MSO}_*(S^k)$ is then free on one generator, α.

For pairs (X,A) and (Y,B) there is a homomorphism

$$\chi : MSO_p(X,A) \otimes MSO_q(Y,B) \to MSO_{p+q}((X,A) \times (Y,B)).$$

By $(X,A) \times (Y,B)$ we mean the pair $(X \times Y, A \times Y \cup X \times B)$. This product is defined by

$$\chi([B^p,f] \otimes [C^q,g]) = [B^p \times C^q, f \times g]$$

where $B^p \times C^q$ has the product orientation. If Y is a point and B is empty we obtain

$$\chi : MSO_p(X,A) \otimes MSO_q \to MSO_{p+q}(X,A)$$

which is just the MSO_*-module structure on $MSO_*(X,A)$ already introduced.

We have vaguely alluded to an analogy between oriented singular bordism and integral singular homology. To be precise there is a canonical relationship given by the *Thom homomorphism*

$$\mu : MSO_n(X,A) \to H_n(X,A;Z).$$

Given a $[B^n,f] \in MSO_n(X,A)$ there is by definition a map

$$f : B^n, \dot{B}^n \to X,A$$

an orientation class $\sigma \in H_n(B^n,\dot{B}^n;Z)$. We set

$$\mu([B^n,f]) = f_*(\sigma) \in H_n(X,A;Z).$$

This is easily seen to be a well defined homomorphism. The image of μ is the subgroup of integral homology classes which are *representable* in the sense of Steenrod. It should be recalled that in [E] Steenrod raised this question: given an integral homology class γ on a complex X then is there a map of a closed oriented manifold in X which carries the orientation class into γ? (We should note here that Thom showed this is not always the case $[T_2]$).

The diagrams

$$\begin{array}{ccc} MSO_n(X,A) & \xrightarrow{\mu} & H_n(X,A) \\ \downarrow{\varphi_*} & & \downarrow{\varphi_*} \\ MSO_n(X_1,A_1) & \xrightarrow{\mu} & H_n(X_1,A_1) \end{array}$$

and

$$\begin{array}{ccc} MSO_n(X,A) & \xrightarrow{\mu} & H_n(X,A) \\ \downarrow{\partial} & & \downarrow{\partial} \\ MSO_{n-1}(A) & \xrightarrow{\mu} & H_{n-1}(A) \end{array} \tag{6.1}$$

both commute. Suppose we set $MSO_+ = \sum_{n>0} MSO_n$, defining thereby the idea of elements of positive degree. The kernel of the Thom homomorphism $\mu : MSO_*(X,A) \to H_*(X,A)$ is then easily seen to contain the image of the natural homomorphism

$$MSO_*(X,A) \underset{MOS*}{\otimes} MSO_+ \to MSO_*(X,A).$$

That is, μ will vanish on bordism classes which are decomposable in the module structure.

Now just as in Eilenberg-Steenrod [ES] there are isomorphisms

$$MSO_r(I^n,S^{n-1}) \simeq MSO_{r-1}(I^{n-1},S^{n-2}),$$

and hence from the commutative diagram

$$\begin{array}{ccc} MSO_n(I^n,S^{n-1}) & \xrightarrow{\mu} & H_n(I^n,S^{n-1}) \\ \downarrow & & \downarrow \\ MSO_{n-1}(I^{n-1},S^{n-2}) & \xrightarrow{\mu} & H_{n-1}(I^{n-1},S^{n-2}) \end{array}$$

we may conclude by induction over n that $\mu : MSO_n(I^n,S^{n-1}) \simeq H_n(I^n,S^{n-1})$ because this is clearly true if $n = 0$. By the relative homeomorphism and the direct sum theorems it follows for any CW-complex X that

$$\mu : MSO_n(X^n,X^{n-1}) \simeq H_n(X^n,X^{n-1})$$

where X^n denotes the n-skeleton. Moreover commutativity holds in

$$(6.2)\quad \begin{array}{ccccc} MSO_n(X^n,X^{n-1}) & \simeq & H_n(X^n,X^{n-1}) & = & C_n(X) \\ \downarrow \partial & & \downarrow \partial & & \downarrow \partial \\ MSO_{n-1}(X^{n-1},X^{n-2}) & \simeq & H_{n-1}(X^{n-1},X^{n-2}) & = & C_{n-1}(X). \end{array}$$

Next we note

$$(6.3)\quad \chi : MSO_n(I^n,S^{n-1}) \otimes MSO_q \simeq MSO_{n+q}(I^n,S^{n-1}).$$

Again this is shown by induction over n beginning with $n = o$ and appealing to

$$\begin{array}{ccc} MSO_n(I^n,S^{n-1}) \otimes MSO_q & \to & MSO_{n+q}(I^n,S^{n-1}) \\ \downarrow & & \downarrow \\ MSO_{n-1}(I^{n-1},S^{n-2}) \otimes MSO_q & \to & MSO_{n+q-1}(I^{n-1},S^{n-2}). \end{array}$$

From this it will also follow that

$$(6.4)\quad \chi : MSO_n(X^n,X^{n-1}) \otimes MSO_q \simeq MSO_{n+q}(X^n,X^{n-1}).$$

We denote by SX the suspension of X. Identifying X with $X \times 1/2$, there are the two cones

$$C_- = (X \times [o,1/2])/X \times o,\ C_+ = (X \times [1/2,1])/X \times 1$$

for which $SX = C_- \cup C_+$ with $X = C_- \cap C_+$. There is a composite isomorphism

$$\Delta : \widetilde{MSO}_n(SX) \simeq MSO_n(SX,C_+) \simeq MSO_n(C_-,X) \simeq \widetilde{MSO}_{n-1}(X).$$

There are also maps $\varphi : X/A \to SA$ of the type well known in stable homotopy as arising from recognizing X/A is homotopy equivalent to X with the cone over A attached. In our case we extend the identity map $A \to A$ to a map $(X,A) \to (C_-,A)$ and let φ be the composite

$$X/A \to C_-/A \simeq SA.$$

(6.5) LEMMA: *The composite homomorphism* $MSO_n(X,A) \simeq \widetilde{MSO}_n(X/A) \xrightarrow{\varphi_*} \widetilde{MSO}_{n-1}(SA) \xrightarrow{\Delta} \widetilde{MSO}_{n-1}(A)$ *agrees with the boundary homomorphism* $\partial : \widetilde{MSO}_n(X,A) \to \widetilde{MSO}_{n-1}(A)$.

PROOF: This follows immediately from the commutative diagram

$$\begin{array}{ccccc} \widetilde{MSO}_n(X/A) & \xleftarrow{\cong} & MSO_n(X,A) & \xrightarrow{\partial} & \widetilde{MSO}_{n-1}(A) \\ \downarrow \varphi_* & & \downarrow \varphi_* & & \downarrow id \\ \widetilde{MSO}_n(SA) & \xleftarrow{\cong} & MSO_n(C_-,A) & \rightarrow & \widetilde{MSO}_{n-1}(A). \end{array} \quad \blacksquare$$

We can also define the suspension isomorphism $S : \widetilde{MSO}_{n-1}(X) \simeq \widetilde{MSO}_n(SX)$ directly. Suppose $[M^{n-1},f]$ represents an element of $\widetilde{MSO}_{n-1}(X)$. Then $M^{n-1} = \dot{B}^n$ for some compact oriented B^n. There is a map F of the boundary $(B^n \times I)^{\cdot}$ into SX which maps $M^{n-1} \times I$ into $X \times I$ via $f \times id$ and which collapses $B^n \times 1$ to the north pole, $B^n \times o$ to the south pole. The resulting $[(B^n \times I)^{\cdot},F]$ in $\widetilde{MSO}_n(SX)$ is uniquely determined since

(6.6) $\Delta[(B^n \times I)^{\cdot},F] = (-1)^{n-1}[M^{n-1},f]$.

We define $S[M^{n-1},f]$ to be $[(B \times I)^{\cdot},F]$ in $\widetilde{MSO}_n(SX)$.

7. The bordism spectral sequence

By elementary considerations we have brought ourselves to the point where the functor $\{MSO_*(X,A),\varphi_*,\partial\}$ undeniably looks like a homology theory. Now it is rather easy to show that if we put $h_n(X,A) = \sum_{p+q=n} H_p(X,A;MSO_q)$ then with the obvious definitions of the homomorphisms we will obtain another generalized homology theory which will agree with oriented bordism on a single point. Indeed $h_*(X,A)$ can be made into an $h_*(pt) = MSO_*$-module if we wish. The most obvious thing to try, then, is to duplicate, on the category of CW-pairs, the Eilenberg-Steenrod uniqueness theorem [ES]. In fact this does not (cannot) work. Instead, in the course of making the attempted generalization there will arise a spectral sequence which represents the relation of $h_*(X,A)$, as the E_2-level, with $MSO_*(X,A)$, as the E_∞-level. Such a spectral sequence is common to generalized homology and cohomology theories and is generally referred to as the *Atiyah-Hirzebruch spectral sequence*.

For a CW-pair (X,A) there are the groups $MSO_p(X^k \cup A, X^j \cup A)$ where X^k is the k-skeleton and there are the exact sequences associated with the triples

$$X^k \cup A \supset X^j \cup A \supset X^m \cup A, \; k \geq j \geq m.$$

Using the terminology of Eilenberg [C_1, Ch.8] there is a spectral sequence $\{E^r_{p,q}\}$ where $E^r_{p,q} = C^r_{p,q}/B^r_{p,q}$ for $C^r_{p,q}$ and $B^r_{p,q}$ the images of

$$MSO_{p+q}(X^p \cup A,\ X^{p-r} \cup A) \to MSO_{p+q}(X^p \cup A,\ X^{p-1} \cup A)$$

$$MSO_{p+q+1}(X^{p+r-1} \cup A,\ X^p \cup A) \to MSO_{p+q}(X^p \cup A,\ X^{p-1} \cup A)$$

respectively. Moreover there is a filtration

$$o \subset J_{o,n} \subset \ldots \subset J_{p,n-p} \subset \ldots \subset J_{n,o} = MSO_n(X,A)$$

with $J_{p,q}$ the image of

$$MSO_{p+q}(X^p \cup A,A) \to MSO_{p+q}(X,A),$$

and $J_{p,q}/J_{p-1,q+1} \simeq E^\infty_{p,q}$.

Now $E^1_{p,q} \simeq MSO_{p+q}(X^p \cup A,\ X^{p-1} \cup A) \simeq C_p(X,A) \otimes MSO_q \simeq C_p(X,A;MSO_q)$ by appealing to (6.4) and (6.2). In addition, $d^1_{p,q} \to E^1_{p-1,q}$ is identified with $\partial : C_p(X,A;MSO_q) \to C_{p-1}(X,A;MSO_q)$ and therefore $E^2_{p,q} \simeq H_p(X,A;MSO_q)$

We have thus for each CW-pair a spectral sequence $\{E^r_{p,q},d^r\}$ whose E_∞-term is associated with a filtration of $MSO_*(X,A)$. If $\varphi : (X,A) \to (X_1,A_1)$ is cellular then there is induced a homomorphism of the spectral sequence of (X,A) into that of (X_1,A_1) which is compatible with $\varphi_* : MSO_*(X,A) \to MSO_*(X_1,A_1)$. If $\varphi_o,\varphi_1 : (X,A) \to (X_1,A_1)$ are cellularly homotopic it follows in a standard way that the homomorphisms $E^r_{p,q} \to {}'E^r_{p,q}$ agree for $r \geq 2$. If $\varphi : (X,A) \to (X_1,A_1)$ is any map between CW-pairs then there is a homotopic map φ' which is cellular [W, p. 22]. Define $\varphi_* : E^r_{p,q} \to {}'E^r_{p,q}$ to be φ'_*. This is well defined for $r \geq 2$ since homotopic cellular maps are cellularly homotopic [W]. In particular, the bordism spectral sequence is independent of the cellular structure for $r \geq 2$.

We should also note that the Thom ring MSO_* acts on the spectral sequence. Each $MSO_*(X^k \cup A,\ X^j \cup A)$ is an MSO_*-module and the inclusion and boundary homomorphisms are MSO_*-module homomorphisms, hence $E^r = E^r_{p,q}$ is an MSO_*-module, each differential d^r is an MSO_*-module homomorphism and $E^{r+1} \simeq H(E^r)$ is an MSO_*-isomorphism. Finally, the product

$$\chi : MSO_*(X,A) \otimes MSO_* \to MSO_*(X,A)$$

has

$$\chi(J_{p,q} \otimes MSO_s) \subset J_{p,q+s}.$$

The induced

$$\chi' : (J_{p,q}/J_{p-1,q+1}) \otimes MSO_s \to J_{p,q+s}/J_{p-1,q+s+1}$$

is the homomorphism $E^\infty_{p,q} \otimes MSO_s \to E^\infty_{p,q+s}$.

(7.1) LEMMA: *The product* $E^2_{p,q} \otimes MSO_s \to E^2_{p,q+s}$ *in the spectral sequence may be identified with the composite homomorphism*

$$H_p(X,A;MSO_q) \otimes MSO_s \to H_p(X,A;MSO_q \otimes MSO_s) \to H_p(X,A;MSO_{q+s}).$$

PROOF: The remark follows from the commutative diagram

$$\begin{array}{ccc} E^1_{p,q} \otimes MSO_s & \longrightarrow & E^1_{p,q+s} \\ \downarrow \simeq & & \downarrow \simeq \\ C_p(X,A;MSO_q) \otimes MSO_s \to C_p(X,A;MSO_q \otimes MSO_s) & \to & C_p(X,A;MSO_{q+s}). \end{array}$$ ∎

(7.2) LEMMA: *The edge homomorphism* $MSO_n(X,A) = J_{n,o} \to E^\infty_{n,o} \to E^2_{n,o} = H_n(X,A;Z)$ *of the bordism spectral sequence coincides with the Thom homomorphism* $\mu : MSO_n(X,A) \to H_n(X,A;Z)$.

PROOF: The edge homomorphism may be described as follows. If $b \in MSO_n(X,A) = J_{n,o}$ then b is the image of an element, c, under $MSO_n(X^n \cup A,A) \to MSO_n(X,A)$. There is also the image, c', of c under $MSO_n(X^n \cup A,A) \to MSO_n(X^n \cup A,X^{n-1})$. Then $c' \in E^1_{n,o} = C_n(X,A)$ and c' represents an element $d \in E^2_{n,o}$. The edge homomorphism maps b into d.

We can now prove (7.2). Consider an element $[B^n,f] \in MSO_n(X,A)$. There is also $[B^n,id] \in MSO_n(B^n,\dot{B}^n)$ and $f_*[B^n,id] = [B^n,f]$. Naturality shows it suffices to prove (7.2) for $[B^n,id] \in MSO_n(B^n,\dot{B}^n)$. Under the edge homomorphism we must show that $[B^n,id]$ maps into the orientation class of B^n. Letting $b = [B^n,id] \in MSO_n(B^n,\dot{B}^n)$ we have c = b. Hence c' is $[B^n,id] \in MSO_n(B^n,X^{n-1})$, where X^{n-1} is the (n-1)-skeleton of B^n. Under the identification

$$\mu : MSO_n(B^n,X^{n-1}) \simeq H_n(B^n,X^{n-1}) = C_n(B^n,\dot{B}^{n-1}),$$

c' is identified with the orientation cycle by definition of μ. The assertion follows. ∎

8. Unoriented bordism

We sketch here the unoriented version $MO_n(X,A)$ of the oriented bordism group $MSO_n(X,A)$. As one would guess from the work of Thom, $MO_n(X,A)$ is much easier to deal with than $MSO_n(X,A)$; in fact the MO_*-module $MO_*(X,A)$ is not difficult to determine completely.

A singular manifold in (X,A) is a map $f : B^n, \dot{B}^n \to X,A$ wherein B^n is a compact manifold. A bordism relation is defined just as in Section 4, except that no orientability requirements are imposed. The resulting group of bordism classes is denoted by $MO_n(X,A)$ and every element in this group has order two. The unoriented bordism class of (B^n,f) is denoted by $[B^n,f]_2$ in $MO_n(X,A)$. The direct sum $MO_*(X,A) = \sum MO_n(X,A)$ is a graded MO_*-module. The functor $\{MO_n(X,A), \varphi_*, \partial\}$ is defined on the category of topological pairs and maps. The first six axioms for a homology are satisfied, however, for a point $MO_n(pt) = MO_n$, the unoriented Thom group.

There is a natural Thom homomorphism $\mu: MO_n(X,A) \to H_n(X,A;Z/2Z)$ defined just as in Section 6. Moreover, for every CW-pair (X,A) there is an unoriented bordism spectral sequence $\left\{E^r_{p,q}, d^r\right\}$ with $E^2_{p,q} \simeq H_p(X,A;MO_q)$ and whose E^∞-term is associated with a filtration of $MO_*(X,A)$. In the unoriented case we see that $H_p(X,A;MO_q) \simeq H_p(X,A;Z/2Z) \otimes MO_q$ and that in fact $\chi: E^2_{p,o} \otimes MO_q \simeq E^2_{p,q}$ for all p,q.

(8.1) THOM: *For any* CW-*pair* (X,A), $\mu : MO_n(X,A) \to H_n(X,A;Z/2Z)$ *is an epimorphism for all* $n \geq o$. ∎

This was shown by Thom in $[T_2]$, and we shall now discuss its implications for unoriented bordism. In the unoriented bordism spectral sequence of (X,A) it follows from the analogue of (7.2) that $E^2_{p,o}$ consists entirely of permanent cycles. Thus $d^2_{p,o} = o$, but since $E^2_{p,q} \simeq E^2_{p,o} \otimes MO_q$ it follows that $d^2_{p,q} = o$, and therefore $E^3_{p,q} = E^2_{p,q}$. Continuing in this manner we learn that $d^r_{p,q} = o$ for all $r \geq 2$. Thus as a corollary to (8.1) we may state

(8.2) COROLLARY: *For every* CW-*pair* (X,A) *the unoriented bordism spectral sequence collapses.* ∎

That is, if $MO_n(X,A) = J_{n,o} \supset \ldots \supset J_{o,n}$ is the filtration coming from the spectral sequence then $J_{n-q,q}/J_{n-q-1,q+1} \simeq E_{n-q,q} \simeq H_{n-q}(X,A;MO_q) \simeq H_{n-q}(X,A;Z/2Z) \otimes MO_q$. Since every element in $MO_n(X,A)$ has order two we see that

$$MO_n(X,A) \simeq \sum H_{n-q}(X,A;Z/2Z) \otimes MO_q.$$

This may be refined into

(8.3) THEOREM: *For any* CW-*pair*, $MO_*(X,A)$ *is a free graded* MO_*-*module isomorphic to* $H_*(X,A;Z/2Z) \otimes MO_*$.

PROOF: We shall not give a complete proof now but it will be completed later. Let $\{c_{n,i}\} \subset H_*(X,A;Z/2Z)$ be a homogeneous basis for the vector space over $Z/2Z$. For each $c_{n,i}$ select an unoriented singular manifold (B_i^n,f_i) in (X,A) with $\mu[B_i^n,f_i]_2 = c_{n,i}$. The elements $\{[B_i^n,f_i]_2\}$ now form a homogeneous MO_*-module basis for $MO_*(X,A)$. The reader can use (8.2) to show $\{[B_i^n,f_i]_2\}$ is a module generating set for $MO_*(X,A)$. Eventually we shall show in passing the independence of this basis. ∎

9. Differentiable bordism groups

Let X^k be a differentiable manifold without boundary about which no assumptions of compactness or orientability are made. We may define *differentiable bordism groups* $D_n(X^k)$. To do so consider pairs (M^n,f) consisting of a closed oriented manifold M^n and a smooth map $f : M^n \to X^k$. Such a pair orientably bords if and only if there is a compact oriented manifold B^{n+1} with $\dot{B}^{n+1} = M^n$ and a smooth map $g : B^{n+1} \to X^k$ with $g|\dot{B}^{n+1} = f$ and such that there is an open set $U \supset \dot{B}^{n+1}$ and a diffeomorphism $h : \dot{B}^{n+1} \times [o,1) \to U$ with $h(x,o) = x$ and with $g(h(x,t)) = f(x)$ for all $o \leq t < 1$ and $x \in \dot{B}^{n+1}$. This last condition is seen to guarantee the transitivity of the bordism relation. The resulting group of bordism classes is denoted by $D_n(X^k)$.

Obviously there is a natural homomorphism $D_n(X^k) \to MSO_n(X^k)$. This is an epimorphism for given a continuous $f : M^n \to X^k$ there is a smooth $f' : M^n \to X^k$ which is homotopic to f. We must now show that $D_n(X^k) \to MSO_n(X^k)$ is a monomorphism. Suppose $f : M^n \to X^k$ is smooth and that there is a compact oriented B^{n+1} with $\dot{B}^{n+1} = M^n$ and a map $g : B^{n+1} \to X^k$ with $g|\dot{B}^{n+1} = f$. Choose an open neighborhood $V \supset \dot{B}^{n+1}$ which is diffeomorphic to $\dot{B}^{n+1} \times [o,1)$ and identify the two. The map $g : \dot{B}^{n+1} \times [o,1/2] \to X^k$ is seen to be homotopic to the map $g' : B^{n+1} \times [o,1/2] \to X^k$ given by $g'(x,t) = f(x)$. By the homotopy extension theorem there is a map $g' : B^{n+1} \to X^k$ with $g'(x,t) = f(x)$ for $o \leq t \leq 1/2$. Using the approximation theorem for smooth maps [M_1, St_2], there is an ε-approximation $G : B^{n+1} \to X^k$ to g' with G smooth and $G = g'$ on $\dot{B}^{n+1} \times [o,1/3]$. Thus f differentiably bords, and we receive the following result

(9.1) THEOREM: *If* X^k *is a smooth manifold without boundary then* $D_n(X^k) \simeq MSO_n(X^k)$. ∎

Similar remarks apply to unoriented bordism. Actually we seldom make an explicit reference to (9.1) but it is often used in the next two chapters so we do ask that our reader keep it in mind.

10. Differential topology and characteristic classes

This will necessarily be an abbreviated treatment of these topics. First we need a few geometric facts to enable us to interpret bordism groups as homotopy groups via the Pontrjagin-Thom construction. For references we might suggest [St_2]. Let us begin by stating an approximation lemma.

(10.1) LEMMA: *Let* $f : M \to N$ *be a continuous map between smooth manifolds without boundary which is differentiable on some closed subset* $A \subset M$. *Let a positive real valued function* $\varepsilon : M \to R^+$ *be given and let* N *have a metric determined by an embedding* $N \subset R^q$. *There exists then a smooth* $g : M \to N$ *which is an* ε*-approximation to* f *and for which* $f|A = g|A$. ∎

We remind the reader that differentiable on A means that at every $x \in A$ there is an open $V_x \subset M$ to which $f|V_x \cap A$ can be extended as a differentiable function.

Continuing for a moment without a boundary we shall, for such a manifold M, denote by M_x the linear space of tangent vectors at $x \in M$. If $f : M \to N$ is smooth then for each $x \in M$ there is the linear transformation $df : M_x \to N_{f(x)}$ which is the differential of f.

The map f is an immersion if and only if df is a monomorphism at each point of M; an embedding if and only if in addition f is a homeomorphism of M onto $im(f) \subset N$. There is now the Whitney embedding theorem [Wy].

(10.2) WHITNEY EMBEDDING THEOREM: *If* $m > 2n$ *then any map of the differentiable manifold* M^n *into* R^m *can be* ε*-approximated by an embedding* g. *If* f *is already an embedding on a neighborhood of the closed set* $A \subset M$ *then we may choose* g *so that* $g|A = f|A$. ∎

We shall also consider manifolds B^n with boundary. Denote by U a neighborhood of $\dot{B}^n$ which is diffeomorphic to $\dot{B}^n \times [o,1)$ and identify U with $\dot{B}^n \times [o,1)$. A map h of B^n into the solid disk D^m is an embedding if and only if h is a 1-1 immersion with $h(\dot{B}^n) \subset S^{m-1} = \dot{D}^m$, $h(B^n) \cap S^{m-1} = h(\dot{B}^n)$

and there exists t_o, $o < t_o < 1$ with $h(x,t) = (1-t)h(x)$ for $(x,t) \in \dot{B}^n \times [o,t_o)$.

(10.3) LEMMA: *If* B^{n+1} *is a compact* $(n+1)$*-manifold and if* $m \geq 2n+2$ *then every embedding of* $\dot{B}^{n+1}$ *into* S^m *can be extended to an embedding of* B^{n+1} *into the solid* $(m+1)$*-disk* D^{m+1}.

PROOF: As usual let U be a neighborhood of $\dot{B}^{n+1}$ which is identified with $\dot{B}^{n+1} \times [o,1)$. Let $h : \dot{B}^{n+1} \to S^m$ be an embedding. Define $h' : B^{n+1} \to D^{m+1}$ by $h'(x,t) = (1-t)h(x)$ for $(x,t) \in B^{n+1} \times [o,1)$ and $h' = o$ otherwise. Now we apply the Whitney embedding theorem to the manifold $B^{n+1} \setminus (\dot{B}^{n+1} \times [o,1/3])$, requiring that the approximation be an extension of h' on $\dot{B}^{n+1} \times (1/3,2/3]$. This yields an embedding of B^{n+1} into D^{m+1} extending h. ■

Let us briefly take up transverse regularity which was extensively developed by Thom to treat cobordism. Suppose N is a smooth manifold without boundary and that $N' \subset N$ is a regularly embedded submanifold. The tangent space N'_x can be regarded as a subspace of the tangent space N_x at each $x \in N'$. The space of normal vectors to N' is by definition the quotient space N_x/N'_x.

Suppose $f : M \to N$ now is a smooth map of a differentiable manifold into N. We say that f is *transverse regular* on N' if and only if for $x \in f^{-1}(N')$ the composite linear transformation

$$M_x \xrightarrow{df} N_{f(x)} \longrightarrow N_{f(x)}/N'_{f(x)}$$

is an epimorphism. Assuming it is non-empty, it is then the case that $f^{-1}(N') \in M$ is a regularly embedded submanifold with

$$\dim M - \dim f^{-1}(N') = \dim N - \dim N'.$$

Furthermore,

$$f|f^{-1}(N') \to N'$$

pulls back the normal bundle to $f^{-1}(N') \subset M$ from the normal bundle to $N' \subset N$. There is a basic approximation theorem [T_2, St_2].

(10.4) THOM: *Let* $f : M^n \to N^m$ *be a smooth map and let* N_1^{m-k} *be a smooth closed submanifold of* N. *Let* A *be a closed subset of* M *such that the transverse regularity condition for* f *and* N_1 *holds at every point in* $A \cap f^{-1}(N_1)$. *Let* δ *be a positive real-valued continuous function on* M. *There exists a smooth map* $g : M^n \to N^m$ *such that*

(1) g *is a* δ*-approximation to* f

(2) g *is transverse regular on* N_1

(3) $g|A = f|A$. ∎

Another item we shall need is *tubular neighborhoods*. Let M^m be a closed manifold. There is a Riemannian metric on M, [S], which we assume fixed. The tangent bundle to M thus receives a smooth innerproduct. Let $V^n \subset M^m$ be a smooth closed submanifold. The bundle $\tau : E \to V^n$ induced by the tangent bundle to M^m splits as a Whitney sum $\tau = \tau_1 \oplus \tau_2$ where $\tau_1 : E_1 \to V^n$ is the tangent bundle to V^n and $\tau_2 : E_2 \to V^n$ is the orthogonal complement of E_1 in E. The bundle τ_2 is naturally equivalent to the normal bundle to V^n in M^m and we identify the two.

Following the classical procedure we define $h : E_2 \to M^m$. A normal vector v at $x \in V^n$ has a length $\|v\|$. If $\|v\| \neq o$ there is a unique geodesic u(s) in M^m, parameterized by arc-length, with u(o) = x and with initial direction $v/\|v\|$. Then $h : E_2 \to M^m$ is defined by

$$h(v) = x \text{ if } v = o$$

$$h(v) = u(\|v\|),\ v \neq o.$$

Identifying V^n with the zero cross-section of E_2, it is then a standard fact that the Jacobian of h is non-singular along $V^n \subset E_2$. Since $h|V^n$ is a diffeomorphism of the compact space and the Jacobian of h is non-singular along V^n there is an open set W, $V^n \subset W \subset E_2$, such that $h : W \to M^m$ is a diffeomorphism onto an open subset of M^m which contains V^n.

Select $\varepsilon > o$ so that if $\|v\| \leq \varepsilon$ then $v \in W$. Let $\eta : D \to V^n$ denote the closed unit cell bundle in E_2; that is, $v \in D$ if and only if $\|v\| \leq 1$. There is then a diffeomorphism $h' : D \to M^m$ onto a compact submanifold of M^m given by $h'(v) = h(\varepsilon v)$. Call the image h'(D) a tubular neighborhood of V^n in M^m with radius ε.

Now there are at least three approaches to introducing characteristic classes associated to a vector bundle. Their interpretation as obstruction classes is discussed in [S]. An axiomatic treatment leading to construction in the fashion of Grothendieck may be found in [H]. We shall very briefy take up now characteristic classes in terms of classifying spaces and their cohomology groups following [B_1, B_2, BH]. Recall that for G a compact Lie group a universal G-bundle is a principal G-bundle $\tau : E(G) \to B(G)$ with E(G) pathwise connected and $\pi_i(E(G)) = o$, $1 \leq i < \infty$ and with B(G) a CW-complex. Call B(G) a classifying space for G. It is homotopically unique. If G is a finite group, then B(G) is K(G,1). For an inclusion $G_1 \subset G_2$ there is associated a fibre map $B(G_1) \to B(G_2)$ with fibre G_2/G_1 which results in a homomorphism

$$\rho = \rho(G_1,G_2) : H^*(B(G_2)) \to H^*(B(G_1)).$$

The idea is that universal characteristic classes associated to G-bundles are to be defined in the cohomology of the classifying space B(G).

First we consider the orthogonal group O(n). In O(n) there is the subgroup D of diagonal matrices. Each diagonal entry must be +1 or -1 so we may think of $D \simeq (C_2)^n$ the n-fold direct product of the cyclic group of order 2 with itself. The inclusion $D \subset O(n)$ then induces a homomorphism

$$\rho : H^*(BO(n);Z/2Z) \to H^*(B(D);Z/2Z).$$

We must describe the image. Now $B(C_2) = K(C_2,1)$ is regarded as $RP(\infty) = \cup_0^\infty RP(n)$, the infinite dimensional real projective space. Thus $H^*(B(C_2);Z/2Z) \simeq Z/2Z[t]$ the polynomial ring over Z/2Z on a single 1-dimensional generator. Thus we can take $B((C_2)^n)$ to be the n-fold cartesian product of $B(C_2)$ with itself. Hence $H^*(B(D);Z/2Z)$ is $Z/2Z[t_1,\ldots,t_n]$ where each t_i is 1-dimensional.

Borel [B_1] has shown that the homomorphism

$$\rho : H^*(BO(n);Z/2Z) \to H^*(B(D);Z/2Z)$$

induced by the inclusion $D \subset O(n)$ is a monomorphism whose image is the subring of symmetric polynomials in $t_1,\ldots,t_n$. Thus there are unique cohomology classes $w_k \in H^k(BO(n);Z/2Z)$, $o \leq k \leq n$, determined by the condition that $\rho(w_k) \in Z/2Z[t_1,\ldots,t_n]$ be the k-th elementary symmetric function. Of course $w_o = 1 \in H^o(BO(n);Z/2Z)$. These w_k are the universal Whitney classes and $H^*(BO(n);Z/2Z) \simeq Z/2Z[w_1,\ldots,w_n]$. An O(n)-bundle $\xi \to X$ over a CW-complex is induced by a homotopically unique map $f : X \to BO(n)$. The image of w_k under

$$f^* : H^k(BO(n);Z/2Z) \to H^k(X;Z/2Z)$$

is the k-th Whitney class of $\xi \to X$ and will also be denoted by w_k. This definition satisfies the axioms for characteristic classes with Z/2Z-coefficients, [H].

From the algebraic fact that every symmetric polynomial in $t_1,\ldots,t_n$ can be expressed as a polynomial in the elementary symmetric functions, it follows that $H^*(BO(n);Z/2Z)$ is the polynomial ring $Z/2Z[w_1,\ldots,w_n]$. An extremely important example is the symmetric function

$$s_k = t_1^k + \ldots + t_n^k, \quad 1 \leq k < \infty.$$

This corresponds to a unique polynomial in the $w_1,\ldots,w_n$ and is also denoted by $s_k \in H^k(BO(n);Z/2Z)$.

The inclusion $SO(n) \subset O(n)$ induces an epimorphism

$$\rho : H^*(BO(n);Z/2Z) \to H^*(BSO(n);Z/2Z)$$

whose kernel is generated by w_1. Thus if we put $w_k = \rho(w_k)$, $2 \leq k \leq n$, we can say that $H^*(BSO(n);Z/2Z) \simeq Z/2Z[w_2,\ldots,w_n]$.

The integral Chern classes associated to a unitary bundle may be treated by analogy. The classifying space of the circle group $S^1 = U(1)$ is $CP(\infty)$, the infinite dimensional complex projective space. Hence $H^*(B(S^1);Z) \simeq Z[y]$, the polynomial ring over Z with a single 2-dimensional generator. The classifying space of the toral group $T^n = (S^1)^n$ is then the n-fold product of $B(S^1)$ with itself so that

$$H^*(B(T^n);Z) \simeq Z[y_1,\ldots,y_n]$$

with each y_i of dimension 2. Now $T^n \subset U(n)$, the group of unitary matrices, as the subgroup of diagonal matrices and again following Borel

$$\rho : H^*(BU(n);Z) \to H^*(B(T^n);Z)$$

is a monomorphism whose image is exactly the subring of symmetric polynomials in the $y_1,\ldots,y_n$. The universal Chern classes

$$c_k \in H^{2k}(BU(n);Z), \quad o \leq k \leq n$$

are then defined to agree with the elementary symmetric functions in $y_1,\ldots,y_n$. Using the classifying map, the Chern classes $c_k \in H^{2k}(X;Z)$ are induced from the universal Chern classes for an $U(n)$-bundle, $\xi \to X$. We note, too, that $H^*(BU(n);Z)$ is $Z[c_1,\ldots,c_n]$.

Pontrjagin classes are associated with orthogonal bundles and arise in the following manner. There is an inclusion $O(n) \subset U(n)$ which induces a homomorphism

$$\rho : H^*(BU(n);Z) \to H^*(BO(n);Z).$$

For each i, $o \leq i \leq n$, we may consider the image of the Chern class $\rho(c_i) \in H^{2i}(BO(n);Z)$. Now if i is odd this image is an element of order 2. Define the universal Pontrjagin class $p_j \in H^{4j}(BO(n);Z)$ to be

$$p_j = (-1)^j \rho(c_{2j}).$$

We understand $p_j = 0$ if $j > [n/2]$. Using the classifying map, Pontrjagin classes are defined for an $O(n)$-bundle over a CW-complex. A note of caution should be injected at this point. We remark that in this form the Pontrjagin classes do not satisfy the Whitney sum theorem. To correct this it is necessary to introduce $Z(1/2)$ as the coefficient ring. This is the ring of rational numbers with denominator a power of 2. Under the coefficient homomorphism $H^*(\cdot;Z) \to H^*(\cdot;Z(1/2))$ we can think of the Pontrjagin classes as elements of $H^{4j}(\cdot;Z(1/2))$. In this setting the Whitney sum theorem is valid.

We may also recognize that there is $U(n) \subset O(2n) \subset U(2n)$ and that if a $U(n)$-bundle is regarded as an $O(2n)$-bundle it will have Pontrjagin classes. The Pontrjagin classes of the universal bundle over $BU(n)$ are computable in terms of the universal Chern classes. The formula is

$$\textstyle\sum_0^n (-1)^j p_j = (\sum_0^n c_i)(\sum_0^n (-1)^i c_i).$$

The formula is valid with integral coefficients.

Formally the Pontrjagin classes can be regarded as the elementary symmetric functions in $y_1^2, \ldots, y_n^2$.

A comment may be made on the origin of the Whitney sum formula in this approach. There is the embedding

$$O(n) \times O(m) \subset O(n+m)$$

which identifies (g,h) with the matrix $\begin{pmatrix} g & 0 \\ 0 & h \end{pmatrix}$. This yields a homomorphism

$$\rho : H^*(BO(n+m);Z/2Z) \to H^*(BO(n) \times BO(m);Z/2Z)$$
$$\simeq H^*(BO(n);Z/2Z) \otimes H^*(BO(m);Z/2Z).$$

Under this homomorphism the universal Whitney classes satisfy

$$\rho(w_k) = \sum_{i+j=k} w_i' \otimes w_j''$$

where $0 \leq k \leq n+m$. The sum formula will follow. The case of Chern classes is similar.

11. Thom spaces

In this section we consider the Thom space of an SO(n)-bundle in preparation for the homotopy interpretation of bordism. The category of spaces with base point is convenient here. The objects are pairs (X,x_o) consisting of a space and a base point $x_o \in X$. The maps must preserve the base points. There is $X \vee Y = X \times y_o \cup x_o \times Y \subset X \times Y$ with base point (x_o,y_o). This is a sum operation and amounts to a disjoint union of X and Y with base points identified. The roof or smash product, of pointed spaces, $X \wedge Y$, is $(X \times Y)/X \vee Y$, the cartesian product with $X \vee Y$ collapsed to a point. In pointed spaces the circle S^1 is taken to be $I/\dot{I}$. The suspension of (X,x_o) is then $SX = X \wedge S^1$. This differs slightly from the suspension of X without regard to a base point in that a specific arc joining the north and south poles is also collapsed to a point.

Turn now to the category of SO(n)-bundles and bundle maps. Suppose $\xi \to X$ is such a bundle. There is the associated disk bundle $D(\xi) \to X$ with fibre the closed n-cell D^n and the associated sphere bundle $S(\xi) \subset D(\xi)$ with fibre S^{n-1}. The *Thom space* $T(\xi)$ is the pointed space $D(\xi)/S(\xi)$. Indeed T is a covariant functor from SO(n)-bundles and bundle maps to pointed spaces and maps. If

$$\begin{array}{ccc} \xi & \xrightarrow{F} & \xi_1 \\ \downarrow & & \downarrow \\ X & \xrightarrow{f} & Y \end{array}$$

is a bundle map then $T(f) : T(\xi) \to T(\xi_1)$ is just induced by $F : (D(\xi),S(\xi)) \to (D(\xi_1),S(\xi_1))$.

If X is a CW-complex then so is $T(\xi)$ with no cell of dimension less than n except for the base point. Hence $T(\xi)$ is (n-1)-connected. We are actually thinking of X as a finite complex. It is true that we shall refer to the Thom space of the universal bundle over BSO(n), but this can be approximated by finite complexes. For this reason we make no comment about topologies.

There is the *Thom isomorphism* $[T_1]$

$$\Psi : H^k(X;Z) \simeq \tilde{H}^{k+n}(T(\xi);Z)$$

which can be described as follows. Over each point of X there is an inclusion $(D^n,S^{n-1}) \subset (D(\xi),S(\xi))$. The *Thom class* $U \in H^n(D(\xi),S(\xi);Z)$ is chosen so that at each point of the base space the image of U under $H^n(D(\xi),S(\xi);Z) \to H^n(D^n,S^{n-1};Z)$ will agree with the prescribed orientation of the cell. Now there is a cup-product pairing

$$H^*(D(\xi),S(\xi);Z) \otimes H^*(D(\xi);Z) \to H^*(D(\xi),S(\xi);Z).$$

But $H^*(X;Z) \simeq H^*(D(\xi);Z)$ and $H^*(D(\xi),S(\xi);Z) \simeq \tilde{H}^*(T(\xi);Z)$. Thus in particular we have

$$\tilde{H}^n(T(\xi);Z) \otimes H^k(X;Z) \to \tilde{H}^{n+k}(T(\xi);Z).$$

The Thom isomorphism is just multiplication by this Thom class U. This isomorphism is natural in that if

$$\begin{array}{ccc} \xi & \xrightarrow{F} & \xi_1 \\ \downarrow & & \downarrow \\ X & \xrightarrow{f} & Y \end{array}$$

is a bundle map then

$$\begin{array}{ccc} \tilde{H}^*(T(\xi_1);Z) & \xrightarrow{F^*} & \tilde{H}^*(T(\xi);Z) \\ \simeq \uparrow \Psi_1 & & \simeq \uparrow \Psi \\ H^*(Y;Z) & \xrightarrow{f^*} & H^*(X;Z) \end{array}$$

commutes.

Suppose now that $\xi \times \eta \to X \times Y$ is the external Whitney sum of an SO(n)-bundle $\xi \to X$ and an SO(m)-bundle $\eta \to Y$. Then $\xi \times \eta$ is an SO(n+m)-bundle and $D(\xi \times \eta)$ has fibre $D^n \times D^m = D^{n+m}$. Now

$$S(\xi \times \eta) = (S(\xi) \times D(\eta)) \cup (D(\xi) \times S(\eta))$$

and

$$T(\xi \times \eta) = T(\xi) \wedge T(\eta).$$

In particular if $R \to X$ is the trivial bundle then

$$T(\xi \oplus R) = T(\xi) \wedge S^1 = ST(\xi).$$

Suppose now we have the universal bundle $\eta_k \to BSO(k)$. The Thom space $T(\eta_k)$ is denoted by MSO(k). The classifying map for $\eta_k \oplus R \to BSO(k)$ will now induce a map $MSO(k) \wedge S^1 \to MSO(k+1)$. This yields the *Thom spectrum* MSO. If n is fixed we find that there are homomorphisms

$$\pi_{n+k}(MSO(k)) \overset{S_*}{\to} \pi_{n+k+1}(MSO(k) \wedge S^1) \to \pi_{n+k+1}(MSO(k+1))$$

where S_* is the suspension homomorphism. We define $\pi_n(MSO)$ to be the direct limit of $\pi_{n+k}(MSO(k)) \to \pi_{n+k+1}(MSO(k+1))$. Later on in the section we shall find that $\pi_n(MSO) \simeq \pi_{n+k}(MSO(k))$ for all $k \geq n+2$, which is stability.

The fundamental motivation behind this construction lies in

(11.1) THOM: *There is an isomorphism of* $\pi_n(MSO)$ *with the oriented bordism group* MSO_n. ∎

The proof may be found in [T_2, St_2]. We shall merely describe the isomorphism

$$MSO_n \simeq \pi_{n+k}(MSO(k)) \quad k \geq n+1.$$

Suppose M^n is a closed oriented n-manifold. Using the Whitney embedding theorem we put $M^n \subset S^{n+k}$. Denote by $\eta \to M^n$ the normal k-plane bundle. Assuming S^{n+k} is oriented then this normal bundle can be uniquely oriented to be compatible with the orientations of the tangent bundles to S^{n+k} and M^n. We consider η to be an SO(k)-bundle. By Section 10, $D(\eta)$ can be identified with a closed normal tubular neighborhood of M^n in S^{n+k}. The classifying map of $\eta \to M^n$ induces

$$T(f) : T(\eta) \to MSO(k).$$

There is then the composite map $g : S^{n+k} \to MSO(k)$ which is given by $S^{n+k} \to S^{n+k}/(S^{n+k} \setminus \operatorname{Int} D(\eta)) \simeq T(\eta) \xrightarrow{T(f)} MSO(k)$. Thus g represents an element of $\pi_{n+k}(MSO(k))$. Thom showed that the correspondence $[M^n] \to [g]$ is well defined and an isomorphism. Here the transverse regularity was essential.

We can at least show

(11.2) LEMMA: *If* $k \geq n+2$ *then the composition of suspension with the homomorphism induced by* $SMSO(k) \to MSO(k+1)$ *yields an isomorphism*

$$\pi_{n+k}(MSO(k)) \simeq \pi_{n+k+1}(MSO(k+1)).$$

PROOF: First of all, since MSO(k) is (k-1)-connected we can apply the suspension theorem to see that

$$\pi_q(MSO(k)) \simeq \pi_{q+1}(SMSO(k))$$

for $q \leq 2k-2$. Now we should observe that cap-product with the Thom class will yield the homology Thom isomorphism and indeed a commutative diagram

$$\begin{array}{ccc} \tilde{H}_{j+k+1}(SMSO(k)) & \longrightarrow & \tilde{H}_{j+k+1}(MSO(k+1) \\ \downarrow \simeq & & \downarrow \simeq \\ H_j(BSO(k)) & \longrightarrow & H_j(BSO(k+1)) \end{array}$$

Now the map $BSO(k) \to BSO(k+1)$ which classifies $\eta_k \oplus R \to BSO(k)$ is actually a fibration with fibre $SO(k+1)/SO(k) = S^k$, the k-sphere. From this we can conclude, using the Thom diagram, that

$$H_{q+1}(SMSO(k)) \simeq H_{q+1}(MSO(k+1))$$

for $o \leq q \leq 2k - 1$. Surely now if $n + 2 \leq k$ then $q = n + k \leq 2k - 2$ and $\pi_{n+k}(MSO(k))$ is isomorphic to $\pi_{n+k+1}(MSO(k+1))$ as required. ∎

12. Homotopy interpretation of bordism groups

By demonstrating the isomorphism $MSO_n \simeq \pi_{n+k}(MSO(k))$ for $k \geq n + 2$, Thom opened the way for a complete analysis of the oriented bordism groups. In this section we prepare for the study of the structure of $MSO_n(X,A)$ by showing how Thom's result extends to an isomorphism

$$MSO_n(X,A) \simeq \pi_{n+k}((X/A \wedge MSO(k))$$

for (X,A) a CW-pair. Put otherwise, oriented bordism is the homology of (X,A) with coefficients the MSO-spectrum.

We shall consider first the easier absolute case. Suppose that (M^n,f) is an oriented singular manifold in X. Embed M^n in S^{n+k} with $k \geq n + 2$. There is a tubular neighborhood N of M^n in S^{n+k} and, by Section 10, N can be considered as the oriented normal k-disk bundle $D(\eta) \to M^n$ to M^n. Suppose that the classifying space $BSO(k)$ is chosen to be a countable CW-complex with $D(\eta_k) \to BSO(k)$ the universal oriented k-cell bundle. By a theorem of Whitehead [W] $BSO(k)$ is of the homotopy type of a locally finite CW-complex. Hence we take $BSO(k)$ to be locally finite. Then, for any CW-complex X, the product $X \times BSO(k)$ is also a CW-complex. Now there is the classifying bundle map

$$\begin{array}{ccc} D(\eta) = N & \xrightarrow{G} & D(\eta_k) \\ \eta \downarrow & & \downarrow \eta_k \\ M^n & \xrightarrow{g} & BSO(k) \end{array}$$

and there is also the composite map $f\eta : N \to M^n \to X$. Hence we see a map

$$(f\eta) \times G : (N,\dot{N}) \to (X \times D(\eta_k),\ X \times S(\eta_k)).$$

Using the same notation for maps we induce

$$(f\eta) \times G : N/\dot{N} \to X \times D(\eta_k)/X \times S(\eta_k).$$

This yields in turn a composition

$$S^{n+k} \to S^{n+k}/(S^{n+k} \setminus \operatorname{Int} N) = N/\dot{N} \to X \times D(\eta_k)/X \times S(\eta_k).$$

We denote this composition by $h : S^{n+k} \to X \times D(\eta_k)/X \times S(\eta_k)$.

(12.1) LEMMA: *The homotopy class in* $\pi_{n+k}(X \times D(\eta_k)/X \times S(\eta_k))$ *of the map* h *depends only on the oriented bordism class* $[M^n,f]$ *in* $MSO_n(X)$.

PROOF: Suppose (M_0,f_0) and (M_1,f_1) are oriented singular manifolds in X and that the M_i are disjointly embedded in S^{n+k}. There are then the (disjoint) normal tubular neighborhoods $\eta_i : N_i \to M_i$ in S^{n+k} and bundle maps

$$\begin{array}{ccc} N_i & \xrightarrow{G_i} & D(\eta_k) \\ \downarrow & & \downarrow \\ M_i & \xrightarrow{g_i} & BSO(k) \end{array}$$

which together with

$$(f_i\eta_i) \times g_i : N_i/\dot{N}_i \to X \times D(\eta_k)/X \times S(\eta_k)$$

give rise to the maps

$$h_i : S^{n+k} \to X \times D(\eta_k)/X \times S(\eta_k).$$

Now suppose that $(-M_0,f_0) \sqcup (M_1,f_1)$ bords. There is then a compact oriented manifold B^{n+1} with $\dot{B}^{n+1} = M_1 \sqcup -M_0$ and a map $F : B^{n+1} \to X$ with $F|M_i = f_i$. Consider then the oriented manifold $I \times S^{n+k}$ with M_0 embedded in $0 \times S^{n+k}$ and M_1 in $1 \times S^{n+k}$. As in Section 10, B^{n+1} can be embedded in $I \times S^{n+k}$ with

$$B^{n+1} \cap (0 \times S^{n+k}) = M_0^n$$

$$B^{n+1} \cap (1 \times S^{n+k}) = M_1^n .$$

We can also suppose there is a t_0, $0 < t_0 < 1$ so that if $(0,x) \in B^{n+1} \cap (0 \times S^{n+k}) = M_0^n$ then $(t,x) \in B^{n+1}$ for $0 < t < t_0$, and similarly for points $(1,x) \in B^{n+1} \cap (1 \times S^{n+k}) = M_1^n$. We also choose a product metric on $I \times S^{n+k}$.

For $\varepsilon > 0$ and sufficiently small, there exists a tubular neighborhood N of radius ε around B^{n+1} on $I \times S^{n+k}$. We may suppose N_0 and N_1 were also of radius ε so that

$$N \cap (0 \times S^{n+k}) = N_0,\ N \cap (1 \times S^{n+k}) = N_1.$$

Now N can be identified with the normal k-disk bundle $D(\eta) \to B^{n+1}$ in $I \times S^{n+k}$. Moreover η restricted to M_i is η_i. Hence there exists a bundle map $G : D(\eta) = N \to D(\eta_k)$ with $G = G_i$ on N_i. Also there is $(F\eta \times G) : N \to X \times D(\eta_k)$ inducing an

$$H : I \times S^{n+k} \to X \times D(\eta_k)/X \times S(\eta_k).$$

In addition $H(0,x) = h_0(x)$ and $H(1,x) = h_1(x)$. The result now follows. ∎

We therefore receive a well defined function

$$\tau : MSO_n(X) \to \pi_{n+k}(X \times D(\eta_k)/X \times S(\eta_k))$$

for $k \geq n + 2$.

(12.2) LEMMA: *The function* τ *is a homomorphism.*

PROOF: Consider singular manifolds (M_0^n, f_0) and (M_1^n, f_1) in X. We have only to embed M_0 in the interior of the lower hemisphere of S^{n+k} and M_1^n in the interior of the upper hemisphere to obtain an embedding of the disjoint union $(M_0^n \sqcup M_1^n,\ f_0 \sqcup f_1)$ into S^{n+k}. One proceeds easily through the definitions to the conclusion. ∎ We observe that if X is a point then $\tau : MSO_n(pt) \to \pi_{n+k}(pt \times D(\eta_k)/pt \times S(\eta_k))$ becomes the Thom isomorphism $MSO_n \cong \pi_{n+k}(MSO(k))$ of (11.1). Thus we may say

(12.3) LEMMA: *For* X *a single point,* τ *is an isomorphism.* ∎

If $\varphi : X \to Y$ is a map then there is a commutative diagram

$$\begin{array}{ccc} MSO_n(X) & \xrightarrow{\tau} & \pi_{n+k}(X \times D(\eta_k)/X \times S(\eta_k)) \\ \downarrow \varphi_* & & \downarrow (\varphi, id)_* \\ MSO_n(Y) & \xrightarrow{\tau} & \pi_{n+k}(Y \times D(\eta_k)/Y \times S(\eta_k)). \end{array}$$

By our convention $X/\emptyset = X \sqcup \infty$. Thus

$$(X/\emptyset) \wedge MSO(k) = (X \times D(\eta_k) \sqcup \infty \times D(\eta_k))/(X \times S(\eta_k) \sqcup \infty \times D(\eta_k))$$

$$\cong X \times D(\eta_k)/X \times S(\eta_k).$$

Using this identification we can represent τ in the following form. Given (M^n,f) and an embedding of M^n into S^{n+k}, $k \geq n+2$, there are maps $f\eta : N \to X \subset X/\emptyset$ and $G : N/\dot{N} \to D(\eta_k)/S(\eta_k) = MSO(k)$ giving a map

$$(f\eta) \wedge G : N/\dot{N} \to (X/\emptyset) \wedge MSO(k)$$

Then $\tau[M^n,f]$ is represented by the composition

$$S^{n+k} \to S^{n+k}/(S^{n+k} \setminus (\operatorname{Int} N)) = N/\dot{N} \xrightarrow{(f\eta)\wedge G} (X/\emptyset) \wedge MSO(k).$$

If a base point $x_o \in X$ is chosen we can say

(12.4) LEMMA: *There is a unique homomorphism*

$$\tau : MSO_n(X,x_o) \to \pi_{n+k}(X \wedge MSO(k)),\ k \geq n+2$$

for which commutativity holds in

$$\begin{array}{ccc} MSO_n(X) & \longrightarrow & MSO_n(X,x_o) \\ \downarrow \tau & & \downarrow \tau \\ \pi_{n+k}((X/\emptyset) \wedge MSO(k)) & \longrightarrow & \pi_{n+k}(X \wedge SO(k)). \end{array}$$

PROOF: There is the sequence

$$o \to (x_o \sqcup \infty) \to X \sqcup \infty \to (X \sqcup \infty)/(x_o \sqcup \infty) = X \to o$$

which produces the exact sequence of spaces

$$o \to (x_o \sqcup \infty) \wedge MSO(k) \to (X/\emptyset) \wedge MSO(k) \to X \wedge MSO(k) \to o.$$

Since $MSO(k)$ is a CW-complex with no cells of dimension less than k except the base point, the three smash products in the preceding short exact sequence are also CW-complexes with no cells of dimension less than k save for the base point. However, by the Blakers-Massey theorem [BM] if X and A are $(k-1)$-connected CW-complexes then the natural homomorphism $\pi_i(X,A) \to \pi_i(X/A)$ is an isomorphism for $i \leq 2k-2$. From the ordinary exact homotopy sequence of the pair $((X/\emptyset) \wedge MSO(k),$ $(x_o/\emptyset) \wedge MSO(k))$ we may therefore conclude with the aid of Blakers-Massey that

$$\pi_{n+k}((x_o/\emptyset) \wedge MSO(k)) \to \pi_{n+k}((X/\emptyset) \wedge MSO(k)) \to$$
$$\pi_{n+k}(X \wedge MSO(k)) \to \ldots$$

is also exact. The assertion (12.4) now is obtained from the diagram

$$\begin{array}{ccccccc} o \to MSO_n(\dot{X}) & \longrightarrow & MSO_n(X) & \longrightarrow & MSO_n(X,x_o) \to o \\ \downarrow & & \downarrow & & \downarrow \\ \pi_{n+k}((x_o/\emptyset) \wedge MSO(k)) & \to & \pi_{n+k}((X/\emptyset) \wedge MSO(k)) & \to & \pi_{n+k}(X \wedge MSO(k)). \end{array} \qquad \blacksquare$$

Using the identification $\widetilde{M}SO_n(X) \simeq MSO_n(X,x_o)$ we thus receive a homomorphism

$$\tau : \widetilde{M}SO_n(X) \to \pi_{n+k}(X \wedge MSO(k)).$$

Recall that $S(X \wedge MSO(k)) = X \wedge SMSO(k)$. We also denote by $S : \pi_i(X) \to \pi_{i+1}(SX)$ the suspension homomorphism which to the homotopy class of $f : S^i \to X$ assigns the homotopy class of $f \to id : S^i \wedge S^1 \to S^i \wedge X \wedge S^1 = SX$. We use the map $SMSO(k) \to MSO(k+1)$ from Section 11 in the next remark.

(12.5) LEMMA: Commutativity holds in

$$\begin{array}{ccccc} & & MSO_n(X) & & \\ & \swarrow^{\tau_1} & & \searrow^{\tau_2} & \\ \pi_{n+k}((X/\emptyset) \wedge MSO(k)) & \to & \pi_{n+k+1}((X/\emptyset) \wedge SMSO(k)) & \to & \pi_{n+k+1}((X/\emptyset) \wedge MSO(k+1)) \end{array}$$

PROOF: Suppose that (M^n,f) is an oriented singular n-manifold in X. Embed M in S^{n+k}, let $\eta : N \to M$ denote the normal tubular neighborhood in S^{n+k} so that there is the classifying bundle map

$$\begin{array}{ccc} N & \xrightarrow{G} & D(\eta_k) \\ \downarrow & & \downarrow \\ M^n & \longrightarrow & BSO(k). \end{array}$$

Then the map

$$(f\eta) \times G : (N,\dot{N}) \to (X/\emptyset) \wedge MSO(k)$$

induces

$$h : S^{n+k} \to (X/\emptyset) \wedge MSO(k)$$

which represents $\tau_1[M^n,f]$. Now S^{n+k} is contained in S^{n+k+1} as the points $(x_1,\dots,x_{n+k+1},o = x_{n+k+2})$. There is then the tubular neighborhood $\eta' : N' \to M$ of $M \subset S^{n+k+1}$, an oriented (k+1)-disk bundle. Obviously $\eta' \to M$ is the Whitney sum of η with a trivial line bundle $\eta \oplus R \to M$.

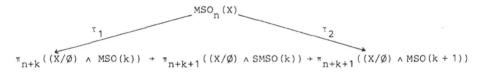

There is then the diagram

$$\begin{array}{ccccc} N' = D(\eta') = D(\eta \oplus R) & \xrightarrow{G \oplus id} & D(\eta_k \oplus R) & \longrightarrow & D(\eta_{k+1}) \\ \downarrow & & \downarrow & & \downarrow \\ M^n & \xrightarrow{\qquad g \qquad} & BSO(k) & \longrightarrow & BSO(k+1) \end{array}$$

Now passing to Thom spaces we get

$$\begin{array}{ccc} N'/\dot{N}' \to D(\eta_k \oplus R)/S(\eta_k \oplus R) & \to & D(\eta_{k+1})/S(\eta_{k+1}) \\ \downarrow & & \downarrow \\ (N/\dot{N}) \wedge S^1 \xrightarrow{G \wedge id} MSO(k) \wedge S^1 & \longrightarrow & MSO(k+1) \end{array}$$

The commutativity in (12.5) follows immediately. ∎

Next there is also an identification

$$S(X \wedge MSO(k)) = (X \wedge S^1) \wedge MSO(k) = (SX \wedge MSO(k))$$

for (X,x_o). Using this we state

(12.6) LEMMA: *There is a commutative diagram:*

$$\begin{array}{ccc} \widetilde{MSO}_n(X) & \xrightarrow{\tau} & \pi_{n+k}(X \wedge MSO(k)) \\ \downarrow & & \downarrow \\ \widetilde{MSO}_{n+1}(SX) & \xrightarrow{\tau} & \pi_{n+k+1}(SX \wedge MSO(k)). \end{array}$$

PROOF: The two vertical arrows both indicate suspension homomorphisms. Suppose that (M^n,f) represents an element of $\widetilde{MSO}_n(X)$. As in Section 6, $S[M^n,f]$ is represented as follows. There is a compact oriented $(n+1)$-manifold B^{n+1} with $\dot{B}^{n+1} = M^n$. Put $W^{n+1} = (B^{n+1} \times I)^{\bullet}$ and define $F : W^{n+1} \to SX$ by $F|M^n \times I = f \times id$, F collapses $B^{n+1} \times o$ and $B^{n+1} \times 1$ into the base point of SX.

Suppose $M^n \subset S^{n+k}$ with normal bundle $\eta \to M$ and $(f\eta) \wedge G : N/\dot{N} \to X \wedge MSO(k)$ induces $h : S^{n+k} \to X \wedge MSO(k)$ representing $\tau[M,f]$.

Using the Whitney embedding theorem as in Section 10 we can embed W^{n+1} into S^{n+k+1} so that $M^n \times [1/2,1] \cup B^{n+1} \times 1$ lies in the upper hemisphere, $M^n \times [o,1/2] \cup B^{n+1} \times o$ lies in the lower hemisphere, and $W^{n+1} \cap S^{n+k} = M^n \times 1/2$, which is identified with M^n. We may also assume that W^{n+1} is orthogonal to S^{n+k} at their points of intersection. Let $\eta' : N' \to W^{n+1}$ be a normal tubular neighborhood of W^{n+1} in S^{n+k+1} chosen so that

$N' \cap S^{n+k} = N$ and so that η' restricted to $M^n \times 1/2$ is $\eta \to M^n$. Now we have maps

$$h : S^{n+k} \to N/\dot{N} \xrightarrow{(f\eta)\wedge G} X \wedge MSO(k)$$

$$h' : S^{n+k+1} \to N'/\dot{N}' \xrightarrow{(F\eta)\wedge G'} SX \wedge MSO(k).$$

We can consider $SX \wedge MSO(k)$ as a union of two cones, say C_+ and C_- with $C_+ \cap C_- = X \wedge MSO(k) \subset SX \wedge MSO(k)$. The map h' maps the upper hemisphere of S^{n+k+1} into C_+, the lower hemisphere into C_- and h' agrees with h on S^{n+k}. It then follows that $[h'] = S[h]$, which is the assertion of (12.6). ∎

We need a stability result generalizing (11.2)

(12.7) LEMMA: *For a CW-pair the composite homomorphism*

$$\pi_{n+k}((X/A) \wedge MSO(k)) \to \pi_{n+k+1}((X/A) \wedge SMSO(k)) \to$$

$$\to \pi_{n+k+1}((X/A) \wedge MSO(k+1))$$

is an isomorphism for $k \geq n + 2$.

PROOF: This is exactly like (11.2). Note that $(X/A) \wedge MSO(k)$ is $(k-1)$-connected. To see that $H_{q+1}((X/A) \wedge SMSO(k);Z) \simeq H_{q+1}((X/A) \wedge MSO(k+1;Z)$ for $o \leq q \leq 2k - 1$ we appeal to the Künneth formula so that for any field, K,

$$\tilde{H}_*((X/A) \wedge MSO(k);K) \simeq H_*((X/A);K) \otimes \tilde{H}_*(MSO(k);K).$$ ∎

With this observation we may define the homology of the pair (X,A) with coefficients in the MSO-spectrum by

$$H_n(X,A;MSO) = \pi_{n+k}((X/A) \wedge MSO(k)), \quad k \geq n + 2.$$

There is a boundary operator

$$\partial : H_n(X,A;MSO) \to H_{n-1}(A,B;MSO)$$

for triples $X \supset A \supset B$ defined by means of the Blakers-Massey theorem. Namely, the natural homomorphism

$$\pi_{n+k}((X/B) \wedge MSO(k),\ (A/B) \wedge MSO(k)) \to \pi_{n+k}((X/A) \wedge MSO(k))$$

is an isomorphism if $k \geq n + 2$ since the spaces involved are $(k-1)$-connected. The boundary operator

$$\partial : \pi_{n+k+1}((X/B) \wedge MSO(k),\ (A/B) \wedge MSO(k)) \rightarrow$$

$$\rightarrow \pi_{n+k}((A/B) \wedge MSO(k))$$

then yields $\partial : H_{n+1}(X,A;MSO) \rightarrow H_n(A,B;MSO)$. Maps $\varphi : (X,A) \rightarrow (X_1,A_1)$ clearly induce homomorphisms $\varphi_* : H_n(X,A;MSO) \rightarrow H_n(X_1,A_1;MSO)$.

(12.8) LEMMA: *The covariant functor* $\{H_*(X,A;MSO),\varphi_*,\partial\}$ *satisfies the first six axioms of Eilenberg-Steenrod for a homology theory.* ∎

We again refer the reader to G.W. Whitehead's study of generalized homology theories [Wh]. There are, as in Section 6, such consequences as reduced groups $\tilde{H}_*(X;MSO)$ and so forth. Here $\tilde{H}_n(X;MSO) \simeq H_n(X,x_o;MSO) \simeq \pi_{n+k}(X \wedge MSO(k))$. There is also a suspension isomorphism $S : \tilde{H}_n(X;MSO) \simeq \tilde{H}_{n+1}(SX;MSO)$, the ordinary suspension

$$\pi_{n+k}(X \wedge MSO(k)) \rightarrow \pi_{n+k+1}(SX \wedge MSO(k)).$$

The composition

$$H_{n+1}(X,A;MSO) \simeq \tilde{H}_{n+1}(X/A;MSO) \xrightarrow{\varphi_*} \tilde{H}_{n+1}(SA;MSO) \rightarrow$$

$$\xrightarrow{S^{-1}} \tilde{H}_n(A;MSO) \rightarrow H_n(A;MSO)$$

agrees with $(-1)^{n+1}\partial$ just as in the case of $MSO_*(X,A)$ (see Section 6).

Now for every CW-pair we have, following (12.4), a homomorphism

$$MSO_n(X,A) \simeq \tilde{M}SO_n(X/A) \rightarrow \tilde{H}_n(X/A;MSO) \simeq H_n(X,A;MSO).$$

The role of (12.5) is to show that τ is well defined; that is, independent of k for k sufficiently large. The role of (12.6), in view of the immediately preceding remarks, is to show that commutativity holds in

$$\begin{array}{ccc} MSO_n(X,A) & \xrightarrow{\tau} & H_n(X,A;MSO) \\ \downarrow \partial & & \downarrow \partial \\ MSO_{n-1}(A) & \xrightarrow{\tau} & H_{n-1}(A;MSO). \end{array}$$

Thus τ is a natural transformation of the oriented bordism into the generalized homology functor $\{H_*(X,A;MSO),\varphi_*,\partial\}$ which on a point induces an isomorphism. This is the pattern for establishing that two generalized homology theories on CW-pairs are isomorphic.

(12.9) THEOREM: *The homomorphism* τ *of* $\{MSO_*(X,A),\varphi_*,\partial\}$ *into* $\{H_*(X,A;MSO),\varphi_*,\partial\}$ *is a natural isomorphism of the two generalized homology theories over the category of* CW-*pairs.*

PROOF: From the Thom result (12.3) τ is an isomorphism for a point; that is, for the coefficient groups. It will follow that τ is an isomorphism for all finite CW-pairs (X,A). Suppose inductively that τ has been shown to be an isomorphism for all pairs (X,A) where X has no more than k cells. Consider then an X with (k+1) cells. If A also has (k+1) cells then X = A and $MSO_*(X,A) = H_*(X,A;MSO) = o$. If A has no more than k cells select a subcomplex X_1, with $A \subset X_1$, and X_1 having exactly k+1 cells. There is then

$$\begin{array}{ccccccc} \cdots \to & MSO_n(X_1,A) & \to & MSO_n(X,A) & \longrightarrow & MSO_n(X,X_1) & \to \cdots \\ & \downarrow \tau_1 & & \downarrow \tau & & \downarrow \tau_2 & \\ \cdots \to & H_n(X,A;MSO) & \to & H_n(X,A;MSO) & \to & H_n(X,X_1;MSO) & \to \cdots \end{array}$$

where τ_1 is an isomorphism by induction. Furthermore, $MSO_n(X,X_1) \simeq MSO_n(I^m,S^{m-1})$ and $H_n(X,X_1;MSO) \simeq H_n(I^m,S^{m-1};MSO)$ for some m. An easy argument using the fact that τ is an isomorphism of coefficient groups shows that τ_2 is also an isomorphism. We apply the five lemma to complete the inductive step.

Having shown that τ is an isomorphism for finite CW-pairs we may proceed to the general case. It is easy to see that for any CW-pair (X,A)

$$MSO_n(X,A) = \text{Dir lim } MSO_n(X_\alpha,A_\alpha)$$

taken over all finite pairs (X_α,A_α) in (X,A).

Similarly

$$\pi_{n+k}((X/A) \wedge MSO(k)) = \text{Dir lim } \pi_{n+k}((X_\alpha/A_\alpha) \wedge MSO(k))$$

so that

$$H_n(X,A;MSO) \simeq \text{Dir lim } H_n(X_\alpha,A_\alpha;MSO).$$

Thus we may conclude that τ is an isomorphism for all CW-pairs. ∎

A *spectrum* for us will be a sequence $\ldots,M_n,M_{n+1},\ldots$ of countable CW-complexes with base points, where M_n is defined for all n sufficiently large, together with maps $SM_n \to M_{n+1}$ such that

a) M_n is (n-1)-connected

b) $SM_n \to M_{n+1}$ induces an isomorphism $\pi_i(SM_n) \simeq \pi_i(M_{n+1})$ for $i < 2n$.

Denote the spectrum by M. For a CW-pair (X,A) we define

$$H_n(X,A;M) = \text{Dir lim } \pi_{n+k}((X/A) \wedge M_k).$$

For X finite dimensional we have

$$H_n(X,A;M) \simeq \pi_{n+k}((X/A) \wedge M), \text{ k large.}$$

There is a boundary operator defined as in the preceding discussion. This always defines a generalized homology theory.

The spectrum M has homotopy groups

$$\pi_n(M) = \text{Dir lim } \pi_{n+k}(M_k)$$

and for an abelian coefficient group K there is

$$H_n(M;K) = \text{Dir lim } \tilde{H}_{n+k}(M_k;K)$$

$$H^n(M;K) = \text{Inv lim } \tilde{H}^{n+k}(M_k;K).$$

There is also

$$\pi_n(M) \simeq H_n(pt;M)$$

by definition. We might observe that for the MSO-spectrum it will follow from the Thom isomorphisms that

$$H^n(MSO;K) \simeq H^n(BSO;K).$$

We caution that this is only an additive isomorphism. Observe carefully that BSO is a space while MSO is a spectrum.

Let us take up another example which will prove important to us later. Fix an abelian group π. There is a spectrum $K(\pi)$ with Eilenberg-MacLane spaces

$$\ldots K(\pi,n),\ K(\pi,n+1),\ldots$$

The maps $SK(\pi,n) \to K(\pi,n+1)$ are given as follows. There are fundamenta cohomology classes $i_m \in H^m(K(\pi,m);\pi)$ and a suspension homomorphism $S : H^m(K(\pi,m);\pi) \to H^{m+1}(SK(\pi,m);\pi)$. There is a homotopically unique map $f : SK(\pi,m) \to K(\pi,m+1)$ with $f^*(i_{m+1}) = S(i_m)$. It is easy to verify conditions a) and b) for a spectrum. According to Cartan-Serre $[C_2]$, the elements of $H^n(K(\pi);K)$ will be the stable cohomology operations

$$H^k(X;\pi) \to H^{k+n}(X;K).$$

Furthermore, the homology groups $H_n(X,A;K(\pi))$ are just $H_n(X,A;\pi)$ by the Eilenberg-Steenrod uniqueness theorem. Simply note that for a point

$$H_n(pt;K(\pi)) \simeq \pi_{n+k}(K(\pi,k))$$

for k large. But $\pi_{n+k}(K(\pi,k)) = o$ if $n > o$ and $\pi_k(K(\pi,k)) \simeq \pi$. Hence $H_*(X,A;K(\pi))$ is the ordinary singular homology theory $H_*(X,A;\pi)$.

13. Duality and cobordism

On the category of finite CW-pairs there is a generalized cohomology theory, dual to bordism, which is called *cobordism* [A]. It is described directly in terms of the Thom spectrum. Although cobordism is not employed here we include a sketch of it for completeness.

We must work in the category of CW-complexes with base point and [X,Y] will denote the set of (based) homotopy classes of base point preserving maps of X into Y. Remember that the sequence of spaces MSO(1),MSO(2),...,MSO(k),... together with the maps $SMSO(k) \to MSO(k+1)$ define the Thom spectrum MSO. For a finite CW-pair the n-th cobordism group is defined by

$$MSO^n(X,A) = \text{Dir lim } [S^k(X/A), MSO(k+n)].$$

The homomorphisms in the indicated direct limit systems are the compositions

$$[S^kX, MSO(n+k)] \xrightarrow{S} [S^{k+1}X, SMSO(n+k)]$$

$$\to [S^{k+1}X, MSO(n+k+1)].$$

(13.1) LEMMA: *If* $\dim X \leq 2n-2$ *then*

$$[X,MSO(n)] \to [SX,MSO(n+1)]$$

is a bijection.

PROOF: Since MSO(n) is (n-1)-connected we can apply the suspension theorem [BM] to see that $[X,MSO(n)] \to [SX,SMSO(n)]$ is a bijection. Now $SMSO(n) \to MSO(n+1)$ is a 2n-equivalence and thus (13.1) follows from [Sp, appendix]. ∎ From this lemma it will follow that

$$MSO^n(X,A) \simeq [S^k(X/A),MSO(k+n)]$$

if $\dim X/A \leq k + 2n - 2$. We also note

(13.2) LEMMA: *If* $n > \dim X/A$ *then* $MSO^n(X,A) = o$.

PROOF: In this case the space $S^k(X/A)$, which has dimension less than $n+k$, is being mapped into $MSO(n+k)$, which is $(n+k-1)$-connected. ∎

The absolute group $MSO^n(X)$ is $MSO^n(X/\emptyset)$. For any map $\varphi : (X,A) \to (Y,B)$ there is an induced homomorphism $\varphi^* : MSO^n(Y,B) \to MSO^n(X,A)$. In the fashion of homotopy there is a coboundary homomorphism $\delta : MSO^n(A) \to MSO^{n+1}(X,A)$. For further details the reader may consult [A].

(13.3) ATIYAH: *On the category of finite* CW-*pairs the cobordism functor* $\{MSO^n(X,A),\varphi^*,\delta\}$ *satisfies the first six Eilenberg-Steenrod axioms for a cohomology theory. For a single point,* $MSO^n(pt) \simeq MSO_{-n}$. ∎

Cobordism, like cohomology, is equipped with a product structure. The cobordism spectral sequence has $E_2^{i,j} \simeq H^i(X,A;MSO^j)$. Since $MSO^j = o$ for $j > o$ it is easy to demonstrate the existence of an edge homomorphism

$$MSO^n(X,A) \to E_2^{n,o} = H^n(X,A;Z).$$

This may be also described directly as the Thom cobordism homomorphism

$$\mu : MSO^n(X,A) \to H^n(X,A;Z).$$

For k sufficiently large an element of $MSO^n(X,A)$ is represented by a map $f : S^k(X/A) \to MSO(k+n)$. This will induce

$$f^* : \tilde{H}^{n+k}(MSO(k+n);Z) \to \tilde{H}^{n+k}(S^k(X/A);Z).$$

Apply f^* to the integral Thom class and define μ to be the element corresponding to $f^*(U_{n+k})$ under the suspension isomorphism

$$H^n(X,A;Z) \simeq \tilde{H}^{n+k}(S^k(X/A);Z).$$

The Thom homomorphism in cobordism is also multiplicative. We turn now to some duality considerations.

(13.4) THOM-ATIYAH: *If* (X,A) *is a finite* CW-*pair for which* $X \setminus A$ *is an oriented* m-*manifold without boundary then there is a canonical isomorphism*

$$u : MSO_n(X \setminus A) \simeq MSO^{m-n}(X,A).$$

PROOF: This will be an outline. Suppose first $n < m/2$. By the embedding theorems then an element of $MSO_n(X \setminus A)$ can be represented by a closed regular submanifold $V^n \subset X \setminus A$. A tubular neighborhood N of V^n in $X \setminus A$

can be identified with the normal (m-n)-disk bundle to V^n, $\eta : D(\eta) \to V^n$. The fibres of N are oriented so that the orientation of the tangent bundle of V followed by the orientation of the normal bundle agrees with the orientation of the tangent bundle to $X \setminus A$ restricted down to $V \subset X \setminus A$. There is then a diagram of bundle maps

$$\begin{array}{ccc} N & \xrightarrow{G} & D(\eta_{m-n}) \\ \downarrow & & \downarrow \\ V & \xrightarrow{g} & BSO(m-n) \end{array}$$

which then induces a map

$$X/A \to N/\dot{N} \to D(\eta_{m-n})/S(\eta_{m-n}) = MSO(m-n).$$

This then determines an element of $[X/A, MSO(m-n)] = MSO^{m-n}(X,A)$. This defines u in the stable range. Also in the stable range commutativity holds in the diagram

$$\begin{array}{ccc} MSO_n(X \setminus A) & \xrightarrow{u} & MSO^{m-n}(X,A) \\ \downarrow i_* & & \downarrow S \\ MSO_n(SX \setminus SA) & \xrightarrow{u} & MSO^{m-n+1}(SX,SA) \end{array}$$

wherein i_* is induced by inclusion.

In arbitrary dimensions therefore the isomorphism u may be defined as the composition $MSO_n(X \setminus A) \xrightarrow[\simeq]{i_*} MSO_n(S^kX \setminus S^kA) \overset{u}{\simeq} MSO^{m-n+k}(S^kX, S^kA) \xleftarrow[\simeq]{} MSO^{m-n}(X,A)$ with k large. ∎

Consider now a finite simplicial complex X embedded as a subcomplex of S^n. For such complexes there is the Spanier-Whitehead duality theory [SpW]. In particular, any finite complex $D_nX \subset S^n \setminus X$ which is a deformation retract of $S^n \setminus X$ is an n-dual of X. If X is a finite CW-complex then there is a finite simplicial complex X' of the same homotopy type as X [W]. An n-dual D_nX' is said to be a weak n-dual, D_nX, of X. As a corollary of (13.4) we can show

(13.5) COROLLARY: *If* X *is a finite* CW-*complex with a weak* n-*dual*, D_nX, *then there is a canonical isomorphism*

$$u' : MSO_k(X) \simeq \widetilde{MSO}^{n-k-1}(D_nX).$$

PROOF: We may confine ourselves to a finite simplicial complex X embedded as a proper subcomplex of S^n. Since X is contractible to a poin in S^n there is a short exact sequence

$$o \to \widetilde{M}SO^{n-k-1}(X) \to MSO^{n-k}(S^n,X) \to \widetilde{M}SO^{n-k}(S^n) \to o.$$

There is also a short exact sequence

$$o \to \widetilde{M}SO_k(S^n \setminus X) \to MSO_k(S^n \setminus X) \to MSO_k \to o.$$

The duality of (13.4) now yields a commutative diagram

$$\begin{array}{ccccccccc} o & \to & \widetilde{M}SO_k(S^n \setminus X) & \to & MSO_k(S^n \setminus X) & \to & MSO_k & \to & o \\ & & & & \downarrow \simeq & & \downarrow \simeq & & \\ o & \to & \widetilde{M}SO^{n-k-1}(X) & \to & MSO^{n-k}(S^n,X) & \to & \widetilde{M}SO^{n-k}(S^n) & \to & o. \end{array}$$

There is therefore a unique isomorphism

$$MSO_k(S^n \setminus X) \simeq \widetilde{M}SO^{n-k-1}(X)$$

which completes the diagram. Since $\widetilde{M}SO_k(S^n \setminus X) \simeq \widetilde{M}SO_k(D_nX)$ we finally obtain the required isomorphism $u' : \widetilde{M}SO_k(D_nX) \simeq \widetilde{M}SO^{n-k-1}(X)$. ∎

14. Triviality mod C

Now we must indicate how, in certain cases, one might actually go about determining the oriented bordism groups $MSO_n(X,A)$. Let us denote by C the class of abelian torsion groups having all elements of odd order [Sr]. Our key result is

(14.1) THEOREM: *For any* CW-*pair* (X,A) *the bordism spectral sequence is trivial mod* C.

To be more explicit, we must show that the image of each $d^r : E^r_{p,q} \to E^r_{p-r,q+r-1}$, with $r \geq 2$, is an odd torsion group. The argument consists in finding a natural transformation of oriented bordism into another generalized homology theory so that, using the proof of (12.9), this natural transformation is a mod C isomorphism and also so that the corresponding Atiyah-Hirzebruch spectral sequences are always trivial. We use in a basic way the following result of C.T.C. Wall [Wa].

(14.2) WALL: *As a module over the mod 2 Steenrod algebra,* $H^*(MSO(k);Z/2Z)$ *is isomorphic in dimensions* $< 2k$ *to a direct sum of Steenrod algebra modules* $H^*(K(Z,m_i);Z/2Z)$ *and* $H^*(K(Z/2Z,n_j);Z/2Z)$. ∎

Because Steenrod operations commute with suspension and with homomorphisms induced by maps we can also regard $H^*(MSO;Z/2Z)$ as a module over the Steenrod algebra. From that viewpoint, $H^*(MSO;Z/2Z)$ is, as a module over the Steenrod algebra, isomorphic to a direct sum of copies of $H^*(K(Z);Z/2Z)$ and $H^*(K(Z/2Z);Z/2Z)$.

Let $a \in H^{m_i}(MSO;Z/2Z)$ be the generator of one of the submodules isomorphic to $H^*(K(Z);Z/2Z)$. The Bockstein $S^1q : H^{m_i}(MSO;Z/2Z) \to H^{m_i+1}(MSO;Z/2Z)$ kills the element a. Hence the integral Bockstein $H^{m_i}(MSO;Z/2Z) \to H^{m_i+1}(MSO;Z)$ carries a into an element of order 2, a', whose reduction mod 2 vanishes. Hence we can write $a' = 2b$ for some $b \in H^{m_i+1}(MSO;Z)$. Using the Thom isomorphism, $H^*(BSO;Z)$ is additively isomorphic to $H^*(MSO;Z)$ and so every torsion class in $H^*(MSO;Z)$ has order 2. From this it follows that $a' = o$ and therefore that a is the reduction mod 2 of an integral cohomology class $\alpha \in H^{m_i}(MSO;Z)$.

For $k > m_i$ these elements, α and a, correspond to unique cohomology classes $\alpha_k \in H^{m_i+k}(MSO(k);Z)$ and $a_k \in H^{m_i+k}(MSO(k);Z/2Z)$. For each $k > m_i$ there is a cellular map

$$f_k : MSO(k) \to K(Z,m_i + k),$$

unique up to homotopy, with $f_k^*(i) = \alpha_k$ where $i \in H^{m_i+k}(K(Z,m_i + k);Z)$ is the fundamental class. Also $f_k^*(i \bmod 2) = a_k$.

The diagram

$$\begin{array}{ccc} SMSO(k) & \longrightarrow & MSO(k+1) \\ \downarrow{\scriptstyle Sf_k} & & \downarrow{\scriptstyle f_{k+1}} \\ SK(Z,m_i + k) & \longrightarrow & K(Z,m_i + k + 1) \end{array}$$

commutes up to homotopy. Put otherwise, we introduce a spectrum $\mathcal{K}(Z,m_i)$ by putting $M_k = K(Z,m_i + k)$ for $k > m_i$ with the obvious connecting maps. The choice of $\alpha \in H^{m_i}(MSO;Z)$ then determines a map of spectra $MSO \to \mathcal{K}(Z,m_i)$. This carries the fundamental class in $H^{m_i}(\mathcal{K}(Z,m_i);Z) \simeq Z$ back into the class $\alpha \in H^{m_i}(MSO;Z)$. Furthermore, if we consider

$$H_s(X,A;\mathcal{K}(Z,m_i)) = \text{Dir lim } \pi_{s+k}((X/A) \wedge \mathcal{K}(Z,m_i + k))$$

we see as in Section 12 that

$$H_s(X,A;K(Z,m_i)) \simeq H_{s-m_i}(X,A;Z).$$

Now there is a generalized Atiyah-Hirzebruch spectral sequence for this homology theory too. But,

$$E^1_{p,q} = H_{p+q}(X^p/X^{p-1};K(Z,m_i)) \simeq H_{p+q-m_i}(X^p/X^{p-1};Z)$$

and so $E^1_{p,q} = o$ if $q \neq m_i$ while $E_{p,m_i} = C_p(X)$. It is also the case that $E^2_{p,m_i} \simeq H_p(X;Z)$. Since there is only one non-zero fibre degree it follows that the differentials d_r, $r \geq 2$, all vanish in this spectral sequence for $H^*(X;K(Z,m_i))$.

The map of spectra $MSO \to K(Z,m_i)$ also yields a natural transformation

$$MSO_*(X,A) \to H_*(X,A;K(Z,m_i))$$

and of course a canonical homomorphism of the bordism spectral sequence into the corresponding (trivial) Atiyah-Hirzebruch spectral sequence.

In Wall's decomposition of $H^*(MSO;Z/2Z)$ consider next a submodule isomorphic to $H^*(K(Z/2Z);Z/2Z)$. Let $b \in H^{n_j}(MSO;Z/2Z)$ be a generator and $b_k \in H^{n_j+k}(MSO(k);Z/2Z)$ a representative. There are maps

$$g_k : MSO(k) \to K(Z/2Z,n_j+k)$$

with $g_k^*(i) = b_k$. This maps MSO into the spectrum $K(Z/2Z,n_j)$. Again

$$H_s(X,A;K(Z/2Z,n_j)) \simeq H_{s-n_j}(X,A;Z/2Z).$$

The Atiyah-Hirzebruch spectral sequence associated with the spectrum $K(Z/2Z,n_j)$ spectrum is trivial too.

We define now a generalized homology theory

$$h_s(X,A) = \sum H_s(X,A;K(Z,m_i)) \oplus \sum H_s(X,A;K(Z/2Z,n_j)).$$

We certainly have a canonical homomorphism

$$\Theta : MSO_*(X,A) \to h_*(X,A)$$

which is a natural transformation of homology theories. There is also a homomorphism of the bordism spectral sequence into the (still trivial) Atiyah-Hirzebruch spectral sequence for $h_*(X,A)$.

We should like to know that Θ is a mod C isomorphism. For a point, $MSO_q(pt) \simeq \pi_{q+k}(MSO(k))$ for k large. Fix this large k and then let

$P_i = K(Z,m_i+k)$ while $Q_j = K(Z/2Z,n_j+k)$. Then

$$h_q(pt) = \sum \pi_{q+k}(K(Z,m_i+k)) \oplus \sum \pi_{q+k}(K(Z/2Z,n_j)) \simeq \pi_{q+k}(\Pi P_i \times \Pi Q_j).$$

The homomorphism $\Theta : MSO_q(pt) \to H_q(pt)$ is induced by the map $f : MSO(k) \to \Pi P_i \times \Pi Q_j$ which is defined by

$$f = \Pi f_k^i \times \Pi g_k^i.$$

According to Wall, $f^* : H^*(\Pi P_j \times \Pi Q_j; Z/2Z) \to H^*(MSO(k); Z/2Z)$ is an isomorphism in dimensions less than 2k. Hence

$$f_* : \pi_{q+k}(MSO(k)) \to \pi_{q+k}(\Pi P_i \times \Pi Q_j)$$

is an isomorphism mod C, the class of odd torsion groups. Thus $\Theta : MSO_q(pt) \to h_q(pt)$ is a mod C isomorphism. From this it follows, just as in (12.9), that Θ is an isomorphism mod C for any CW-pair.

Coming back to the spectral sequences there is the homomorphism $\left\{E^r_{p,q}\right\} \to \left\{'E^r_{p,q}\right\}$ with $\left\{'E^r_{p,q}\right\}$ trivial for $r \geq 2$. Moreover, $\Theta : E^1_{p,q} \to {}'E^1_{p,q}$ is an isomorphism mod C and hence $E^r_{p,q} \to {}'E^r_{p,q}$ is also an isomorphism for all $r \geq 1$. This completes the proof of (14.1). ∎

(14.2) THEOREM: *For any* CW-*pair there is a mod* C *isomorphism*

$$\Theta : MSO_n(X,A) \simeq \sum_{p+q=n} H_p(X,A;MSO_q).$$

For any finite CW-*complex there is a mod* C *isomorphism.*

$$MSO^m(X) \simeq \sum_{p-q=m} H^p(X;MSO_q).$$

PROOF: In the proof of (14.1) we constructed a mod C isomorphism $\Theta : MSO_n(X,A) \simeq H_n(X,A)$, where $h_n(X,A) \simeq \sum H_{n-m_i}(X,A;Z) \oplus \sum H_{n-n_j}(X,A;Z/2Z)$

We can also say that

$$h_n(X,A) = \sum_{p+q=n} H_p(X,A;\Lambda_q)$$

where Λ_q is the direct sum of as many copies of Z as there are i with $m_i = q$ with as many copies of Z/2Z as there are j with $n_j = q$. In particular, $h_n(pt) \simeq \Lambda_n$ and there is the mod C isomorphism $\Theta : MSO_n(pt) \simeq h_n(pt)$ or $\Theta : MSO_n \to \Lambda_n$. But MSO_n is a finitely generated abelian group all of whose torsion elements have order 2. The same is true of Λ_n and when two such groups are mod C isomorphic then they are (abstractly) isomorphic. Hence $\Lambda_q \simeq MSO_q$ and

$$MSO_n(X,A) \simeq \sum_{p+q=n} H_p(X,A;MSO_q)$$

mod C. The result for $MSO^m(X)$ follows by duality arguments which are omitted. ∎

15. Steenrod representability

We return now to the Thom homomorphism $\mu : MSO_n(X,A) \to H_n(X,A;Z)$ introduced in Section 6. To each oriented singular manifold (B^n,f) in (X,A), μ assigns the image of the orientation class of B^n under the induced homomorphism $f_* : H_n(B^n,\dot{B}^n;Z) \to H_n(X,A;Z)$. This is the edge homomorphism of the bordism spectral sequence. The image of μ is the subgroup of integral homology classes which are representable in the sense of Steenrod. We can now make some progress in the study of μ, using the fact (7.2) that the image of μ is the set of permanent cycles of the term $E^2_{n,o}$ in the bordism spectral sequence.

(15.1) THEOREM: *If* (X,A) *is a CW-pair then the bordism spectral sequence collapses (is trivial) if and only if* $\mu : MSO_n(X,A) \to H_n(X,A;Z)$ *is an epimorphism for all* $n \geq o$.

PROOF: By definition the spectral sequence collapses if and only if $d^r : E^r_{p,q} \to E^r_{p-r,q+r-1}$ is trivial for all $r \geq 2$. It is clear that if the spectral sequence collapses, then $\mu : MSO_n(X,A) \to H_n(X,A;Z)$ is an epimorphism for all $n \geq o$.

Let us therefore assume that μ is an epimorphism. Then $d_r : E^r_{n,o} \to E^r_{n-r,r+}$ is trivial for all $r \geq 2$. Consider the operation of MSO_* on the spectral sequence as in Section 7. We have $H_p(X,A;Z) \otimes MSO_q = E^2_{n,o} \otimes MSO_q \to E^2_{p,q} \simeq H_p(X,A;MSO_q)$. From (7.1) we see this is a monomorphism with image

$$H_p(X,A;Z) \otimes MSO_q \subset H_p(X,A;MSO_q).$$

Since every element of $E^2_{p,o}$ is a permanent cycle, so is every element in

$$H_p(X,A;Z) \otimes MSO_q \subset H_p(X,A;MSO_q).$$

Now $H_p(X,A) \otimes MSO_q$ is a direct summand of $E^2_{p,q}$, the other summand, $T_{p,q}$, being isomorphic to $Tor(H_{p-1}(X,A;Z);MSO_q)$. Since MSO_q has no odd torsion, $T_{p,q}$ consists of 2-torsion only. But by (14.1) d^2 carries $E^2_{p,q}$

onto an odd torsion group and hence $d^2(T_{p,q}) = o$ also. So we find $d^2 = o$ and $E^2_{p,q} \simeq E^3_{p,q}$. Continuing on through the spectral sequence it is thus seen to collapse. ∎

It was Thom who first showed μ need not be an epimorphism [T_2]. Specifically $H_7(K(C_3 \times C_3,1);Z)$ contains an element which is not Steenrod representable. The bordism spectral sequence of $K(C_3 \times C_3,1)$ is therefore non-trivial.

(15.2) THEOREM: *If* (X,A) *is a* CW-*pair for which each* $H_n(X,A;Z)$ *is finitely generated and has no odd torsion then the bordism spectral sequence collapses. Moreover* $\mu : MSO_n(X,A) \to H_n(X,A;Z)$ *is an epimorphism and*

$$MSO_n(X,A) \simeq \sum_{p+q=n} H_p(X,A;MSO_q).$$

PROOF: The bordism spectral sequence is trivial mod C and $E^2_{p,q} = H_p(X,A;MSO_q)$ has no odd torsion so that $d^2 = o$ and $E^3_{p,q} = E^2_{p,q}$. We repeat the argument, eventually showing the spectral sequence collapses.

Since $E^2_{p,q}$ has no odd torsion neither does $MSO_*(X,A)$. From (14.2) we have a mod C isomorphism $MSO_n(X,A) \simeq \sum_{p+q=n} H_p(X,A;MSO_q)$ joining finitely generated abelian groups without odd torsion. Hence $MSO_n(X,A) \simeq \sum_{p+q=n} H_p(X,A;MSO_q)$. ∎

(15.3) COROLLARY: *Let* (X,A) *be a* CW-*pair. For each homology class* $c \in H_n(X,A;Z)$ *there is an integer* k *for which* (2k + 1)c *is Steenrod representable.*

PROOF: This is clear since by (14.1) we know that the cokernel of $\mu : MSO_n(X,A) \to H_n(X,A;Z)$ is an odd torsion group. ∎ We also point out as a corollary to (14.1).

(15.4) COROLLARY: *If* (X,A) *is a* CW-*pair then every element in the* 2-*primary torsion subgroup of* $H_n(X,A;Z)$ *is Steenrod representable.* ∎

There is also a gap theorem for the bordism spectral sequence which we examine next.

(15.5) LEMMA: *For a* CW-*pair* (X,A), $E^r_{p,q}$ *consists entirely of elements of order* 2 *if* $q \neq o \pmod 4$.

PROOF: As we noted in Section 2 the results of Milnor and Wall show that MSO_q consists entirely of elements of order 2 if $q \neq o \pmod 4$.

Now

$$E^2_{p,q} \simeq H_p(X,A;Z) \otimes MSO_q \oplus \mathrm{Tor}(H_{p-1}(X,A;Z),MSO_q)$$

implies the result for E^2 and hence for E^r. ∎

(15.6) LEMMA: *The boundary* $d^r : E^r_{p,q} \to E^r_{p-r,q+r-1}$ *is trivial unless* $q \equiv 0$ *(mod 4) and* $r \equiv 1$ *(mod 4).*

PROOF: This follows from (14.1) and (15.5). ∎ We come now to the gap theorem.

(15.7) THEOREM: *Let* (X,A) *be a CW-pair. If for each pair of integers* $r > s$ *with* $r - s \equiv 1$ *(mod 4) either* $H_r(X,A;Z)$ *is a 2-primary torsion group or* $H_s(X,A;Z)$ *has no odd torsion then every element of* $H_n(X,A;Z)$, $n \geq 0$, *is Steenrod representable.*

PROOF: Using (15.6), $E^5_{p,q} \simeq E^2_{p,q} \simeq H_p(X,A) \otimes MSO_q \oplus \mathrm{Tor}(H_{p-1}(X,A),MSO_q)$. Thus if $H_p(X,A)$ is a 2-primary torsion group so is $E^5_{p,q}$ and by (14.1), $d^5 : E^5_{p,q} \to E^5_{p-5,p+4}$ is trivial. If $H_{p-5}(X,A)$ has no odd torsion then neither does $E^5_{p-5,p+4}$ so that $d^5 : E^5_{p,q} \to E^5_{p-5,q+4}$ is again trivial. Thus $E^2_{p,q} \simeq E^5_{p,q} \simeq E^9_{p,q}$ and we repeat the argument inductively to show the bordism spectral sequence collapses. ∎

16. A generalization of Rochlin's theorem

There is the homomorphism $r : MSO_n \to MO_n$ given by ignoring the orientation of the closed manifold; that is, $r[M^n] = [M^n]_2$. It was shown by Rochlin [R] that the sequence

$$MSO_n \xrightarrow{2} MSO_n \xrightarrow{r} MO_n$$

is exact at the middle term. In this section we shall show that for any CW-pair the corresponding sequence

$$MSO_n(X,A) \xrightarrow{2} MSO_n(X,A) \xrightarrow{r} MO_n(X,A)$$

is also exact, where $r[B^n,f] = [B^n,f]_2$.

Let us make a few remarks about unoriented bordism. Let $\eta'_k \to BO(k)$ be the universal orthogonal k-plane bundle. There is a Thom space $MO(k)$ and a Thom spectrum MO. Just as in Section 12,

$$MO_n(X,A) \cong H_n(X,A;MO) \cong \pi_{n+k}((X/A) \wedge MO(k))$$

for k large. Clearly there is a bundle map

$$\begin{array}{ccc} \eta_k & \longrightarrow & \eta'_k \\ \downarrow & & \downarrow \\ BSO(k) & \longrightarrow & BO(k) \end{array}$$

which gives rise to a map $\nu : MSO(k) \to MO(k)$, and hence to a map $MSO \to MO$ of spectra. We leave as an exercise the following

(16.1) LEMMA: *The diagram*

$$\begin{array}{ccc} MSO_n(X,A) & \longrightarrow & \pi_{n+k}((X/A) \wedge MSO(k)) \\ \downarrow r & & \downarrow (id \wedge \nu)_* \\ MO_n(X,A) & \longrightarrow & \pi_{n+k}((X/A) \wedge MO(k)) \end{array}$$

is commutative for k *large.* ∎

We can proceed now to the proof of the theorem

(16.2) THEOREM: *For a* CW-*pair* (X,A) *the sequence*

$$MSO_n(X,A) \xrightarrow{2} MSO_n(X,A) \xrightarrow{r} MO_n(X,A)$$

is exact at the middle term.

PROOF: Since every element of $MO_n(X,A)$ has order 2 we see Im $2 \subset$ Ker(r) Suppose now that $x \in MSO_n(X,A)$ has $r(x) = o$. The idea of the proof is to show that under the mod C isomorphism

$$\Theta : MSO_n(X,A) \to h_n(X,A)$$

of Section 14 the image $\Theta(x)$ is divisible by 2. It then follows that x is already divisible by 2 since C is the class of odd torsion groups. In order to show that the image of x is divisible by 2 it is enough to show that the image of x under each of the homomorphisms

$$MSO_n(X,A) \to H_n(X,A;K(Z,m_i))$$

$$MSO_n(X,A) \to H_n(X,A;K(Z/2Z,n_j)$$

is divisible by 2. We proceed now to show this for the case $MSO_n(X,A) \to H_n(X,A;K(Z,m_i))$.

Recall the definition of $H_n(X,A;K(Z,m_i))$. For k large there was a certain class a_k in $H^{m_i+k}(MSO(k);Z/2Z)$ which is the restriction of an integral class $\alpha_k \in H^{m_i+k}(MSO(k);Z)$. There was then a map $f_k^* : MSO(k) \to K(Z,m_i+k)$ with $f_k^*(i) = \alpha_k$ and f_k^* (i mod 2) $= a_k$. Then

$$\begin{aligned} & H_n(X,A;K(Z,m_i)) \\ \simeq\ & \pi_{n+k}((X/A) \wedge K(Z,m_i+k)) \\ \simeq\ & H_{n-m_i}(X,A;Z). \end{aligned}$$

Now $BSO(k) \to BO(k)$ induces an epimorphism

$$H^*(BO(k);Z/2Z) \to H^*(BSO(k);Z/2Z)$$

so by using the Thom isomorphism for coefficients $Z/2Z$ we conclude that

$$\nu^* : H^*(MO(k);Z/2Z) \to H^*(MSO(k);Z/2Z)$$

is also an epimorphism. Thus there is a $c_k \in H^{m_i+k}(MO(k);Z/2Z)$ for which $\nu^*(c_k) = a_k$. Define $F_k : MO(k) \to K(Z/2Z,m_i+k)$ so that $F_k^*(i') = c_k$ where $i' \in H^{m_i+k}(K(Z/2Z,m_i+k);Z/2Z)$ is the fundamental class.

In terms of spectra we now have

$$\begin{array}{ccc} MSO & \xrightarrow{f} & K(Z,m_i) \\ \downarrow{\scriptstyle \nu} & & \downarrow{\scriptstyle \nu'} \\ MO & \xrightarrow{F} & K(Z/2Z,m_i) \end{array}$$

and hence we have a commutative diagram

$$\begin{array}{ccccc} H_n(X,A;MSO) & \xrightarrow{\Theta_i} & H_n(X,A;K(Z,m_i)) & \simeq & H_{n-m_i}(X,A;Z) \\ \downarrow{\scriptstyle r} & & \downarrow & & \downarrow \\ H_n(X,A;MO) & \xrightarrow{F_*} & H_n(X,A;K(Z/2Z,m_i)) & \simeq & H_{n-m_i}(X,A;Z/2Z) \end{array}$$

Since any element in the kernel of $H_{n-m_i}(X,A;Z) \to H_{n-m_i}(X,A;Z/2Z)$ must lie in the image of $H_{n-m_i}(X,A;Z) \xrightarrow{2} H_{n-m_i}(X,A;Z)$ it will follow that if $r(x) = o$ then $\Theta_i(x)$ is divisible by 2.

On the other hand, if we consider the case

$$\Theta_j : MSO_n(X,A) \to H_n(X,A;K(Z/2Z,n_j)) \simeq H_{n-n_j}(X,A;Z/2Z)$$

we could argue in a similar vein that if $r(x) = o$ then $\Theta_j(x) = o$.

In this manner we find that $r(x) = o$ implies $\Theta(x) \in h_n(X,A)$ is divisible by 2. Since Θ is an isomorphism module odd torsion we may conclude $x \in 2\, MSO_n(X,A)$. ∎

17. Algebraic invariants of maps

In this section we generalize the well known Stiefel-Whitney numbers and Pontrjagin numbers of a closed manifold, obtaining thereby *Whitney numbers* and *Pontrjagin numbers of a map* $f : M^n \to X$. We shall prove theorems showing when we can expect these algebraic invariants to determine the bordism classes.

Let us begin with the mod 2 case. If $\omega = (i_1 \leq i_2 \leq \ldots \leq i_k)$ is a finite sequence of non-negative integers then we put $|\omega| = i_1 + \ldots + i_k$. Associated to each such sequence there is a product

$$V_\omega = v_{i_1} \ldots v_{i_k} \in H^{|\omega|}(BSO;Z/2Z)$$

where $v_1, v_2, \ldots$ denote the universal Whitney classes. Now if B^n is a compact manifold then there is a homotopically unique map $\tau : B^n \to BO(n)$ classifying the tangent bundle. The composition $B^n \to BO(n) \to BO$, which will classify the stable tangent bundle, will also be denoted by τ. We shall put $W_\omega = \tau^*(V_\omega)$ to denote the appropriate product of the tangential Stiefel-Whitney classes. The restriction of $\tau : B^n \to BO$ down to $\dot{B}^n$ stably classifies the tangent bundle to the boundary and so under $i^* : H^{|\omega|}(B;Z/2Z) \to H^{|\omega|}(\dot{B};Z/2Z)$ the tangential class of B is carried into the tangential class of $\dot{B}$.

In the case of a closed manifold the mod 2 characteristic numbers are defined as follows. Let $\sigma_n \in H_n(M^n;Z/2Z)$ be the mod 2 fundamental homology class (mod 2 orientation class). To every sequence ω, with $|\omega| = n$, we associate the characteristic number (Stiefel-Whitney number)

$$\langle W_\omega, \sigma_n \rangle \in Z/2Z.$$

By the symbol $\langle x, \sigma_n \rangle$ for $x \in H^n(M;Z/2Z)$ we mean

$$\varepsilon_*(x \cap \sigma_n) \in H_0(pt;Z/2Z) \simeq Z/2Z,$$

which is just the Kronecker pairing. Pontrjagin showed that if $[M^n]_2 = 0$ then all the Stiefel-Whitney numbers vanish. The proof is as follows. Suppose B^{n+1} is a compact manifold with $\dot{B}^{n+1} = M^n$. If $|\omega| = n$ then there is the product $\widetilde{W}_\omega \in H^n(B^{n+1};Z/2Z)$ of the tangential classes of B^{n+1} and $i^*(\widetilde{W}_\omega) = W_\omega \in H^n(M^n;Z/2Z)$. Now by the well known property of the Kronecker pairing

$$<W_\omega,\sigma_n> = <i^*(W_\omega),\sigma_n> = <\widetilde{W}_\omega, i_*(\sigma_n)> \in Z/2Z.$$

Of course $i_*(\sigma_n) = 0 \in H_n(B^{n+1};Z/2Z)$. Thom showed the converse by proving $[T_2]$

(17.1) THOM: *If* M^n *is a closed manifold then* $[M^n]_2 = 0$ *if and only if all the Stiefel-Whitney numbers of* M^n *vanish.* ∎

Next consider a map $f : M^n \to X$ where M^n is still a closed n-manifold. To every cohomology class $h \in H^m(X;Z/2Z)$ and to every partition ω of n-m there is associated a Stiefel-Whitney number of the form

$$<W_\omega f^*(h),\sigma_n> \in Z/2Z.$$

These are the Stiefel-Whitney numbers of the map f associated with the cohomology class h. If $h = 1 \in H^0(M;Z/2Z)$ is the unit cohomology class then the Stiefel-Whitney numbers of f associated to 1 are clearly the ordinary Stiefel-Whitney numbers of the closed manifold M.

Again it is true that if $[M^n,f]_2 = 0$ then all characteristic numbers of $f : M^n \to X$ vanish. Suppose $F : B^{n+1} \to X$ is a map with $F|\dot{B}^{n+1} = f : M^n \to X$. Then $<W_\omega f^*(h),\sigma_n> = <i^*(\widetilde{W}_\omega F^*(h)),\sigma_n> = <\widetilde{W}_\omega F^*(h), i_*(\sigma_n)> = 0$. Thus the characteristic numbers of $f : M^n \to X$ are seen to depend only on $[M^n,f]_2 \in MO_n(X)$.

Let X be a finite CW-complex. For each $n \geq 0$ let $\{c_{n,i}\}$ denote a basis for $H_n(X;Z/2Z)$ as a vector space over Z/2Z. According to (8.1), for each $c_{n,i}$ we can choose a singular manifold $f_i : M_i^n \to X$ with $f_{i*}(\sigma_n) = c_{n,i}$. We consider the free MO_*-module

$$H_*(X;Z/2Z) \otimes_{Z/2Z} MO_*$$

and define an MO_*-module homomorphism

$$h : H_*(X;Z/2Z) \otimes MO_* \to MO_*(X)$$

by $h(c_{n,i} \otimes 1) = [M_i^n,f_i]_2 \in MO_n(X)$. Since $H_*(X;Z/2Z) \otimes MO_*$ is a free graded MO_*-module, h is well defined.

(17.2) THEOREM: *The* MO_**-module homomorphism* $h : H_*(X;Z/2Z) \otimes MO_* \to MO_*(X)$ *is an isomorphism.*

PROOF: Consider first the Whitney numbers of a product

$$[M_i^n, f_i]_2 [V^m]_2 = [M_i^n \times V^m, f_i\pi]_2$$

where π is the projection $M_i^n \times V^m \to M_i^n$. Let $c^{n,i} \in H^n(X;Z/2Z)$ be the cohomology class dual to $c_{n,i}$; that is, $\langle c^{n,i}, c_{n,i}\rangle = 1$ while $\langle c^{n,i}, c_{n,j}\rangle = o$ for $i \neq j$. Select a partition $\omega = (i_1 \leq \ldots \leq i_k)$ with $|\omega| = m$ and consider the characteristic number associated to $c^{n,i}$ for the composite map $f_i\pi$. Note that $(f_i\pi)^* c^{n,i} = f^*(c^{n,i}) \otimes 1$. Let $v_j \in H^j(V^m;Z/2Z)$ denote the Stiefel-Whitney classes of V^m. Any Stiefel-Whitney class of the product $M_i^n \times V^m$ has the form $w_j = 1 \times v_j$ + terms involving positive dimensional Stiefel-Whitney classes of M_i^n. Thus from dimensional considerations

$$\langle W_\omega (f_i\pi)^*(c^{n,i}), \sigma_n \otimes \sigma_m\rangle =$$

$$\langle f_i^* (c^{n,i}) \otimes V_\omega, \sigma_n \otimes \sigma_m\rangle = \langle f_i^*(c^{n,i})\sigma_n\rangle\langle V_\omega, \sigma_m\rangle =$$

$$\langle c^{n,i}, c_{n,i}\rangle\langle V_\omega, \sigma_n\rangle = \langle V_\omega, \sigma_m\rangle.$$

Furthermore, the characteristic numbers of $f_i\pi$ associated with $c^{n,j}$ with $j \neq i$ will all vanish.

Now suppose there is an expression, n fixed, $\sum_{m,i} [M_i^{n-m}, f_i][V_i^m]_2 = o$. All characteristic numbers must vanish. We shall show inductively that $[V_i^m]_2 = o$, all m,i. Suppose that for $m < m_0$ and all i it has been shown that $[V_i^m] = o$. Choose a partition $\omega = (i_1 \leq \ldots \leq i_k)$ with $|\omega| = m_0$. and a c^{n-m_0, i_0}. Then

$$\sum_{m \geq m_0, i} \langle W_\omega (f_i\pi)^* c^{n-m_o, i_o}, M_i^{n-m} \times V_i^m\rangle = o.$$

Now observe from dimensional considerations that if $m > m_0$ then

$$\langle W_\omega (f_i\pi)^* c^{n-m_o, i_o}, M_i^{n-m} \times V^m\rangle =$$

$$\langle W_\omega (f_i^*(c^{n-m_o, i_o}) \otimes 1, M_i^{n-m} \times V_i^m\rangle = o,$$

simply because $n - m_0 > n - m$. Thus we find

$$\sum_i \langle W_\omega (f_i\pi)^* c^{n-m_o, i_o}, M_i^{n-m_o} \times V_i^{m_o}\rangle = o.$$

However, if $i \neq i_0$ the summand is o, while if $i = i_0$ the summand is the Stiefel-Whitney number $<V_\omega, V_{i_o}^{m_o}>$. Hence all Stiefel-Whitney numbers of $V_{i_0}^{m_0}$ vanish and $[V_{i_0}^{m_0}]_2 = o$. This establishes the inductive step. Thus we have shown h is a monomorphism.

The partial proof given for (8.3) shows h is an epimorphism. This completes (8.3) of course. ∎ In passing we have also established the following

(17.3) THEOREM: *If* $f : M^n \to X$ *is an unoriented singular manifold in a finite* CW-*complex then* $[M^n, f]_2 = o$ *if and only if every characteristic number of the map vanishes.* ∎

It is, incidentally, trivial to see here that all we have used is the assumption that each $H_n(X; Z/2Z)$ is finite dimensional over $Z/2Z$.

Let us treat Whitney numbers in a more functorial fashion. We shall use the stable tangent bundle to a compact manifold. For a compact manifold we denote by $\tau: B^n \to BO$ the homotopically unique map which classifies the stable tangent bundle, and we note that the restriction $\tau|\dot{B}^n \to BO$ still classifies the stable tangent bundle to $\dot{B}^n$.

On the category of finite CW-pairs we can introduce two generalized cohomology functors

$$\hom^n(X,A) = \mathrm{Hom}_{Z/2Z}(MO_n(X,A); Z/2Z)$$

and

$$H^n(BO \times X, BO \times A; Z/2Z) \simeq \sum_{p+q=n} H^p(BO; Z/2Z) \otimes H^q(X,A; Z/2Z).$$

In the first case the induced homomorphisms and the coboundary are all defined by duality. There is a natural transformation

$$T : H^n(BO \times X, BO \times A; Z/2Z) \to \hom^n(X,A).$$

Let $\gamma \in H^n(BO \times X, BO \times A; Z/2Z)$, then a corresponding $T_\gamma \in \hom^n(X,A)$ is given as follows. If $f : (B^n, \dot{B}^n) \to (X,A)$ is a singular manifold then there is the map $\tau \times f : (B^n, \dot{B}^n) \to (BO \times X, BO \times A)$ and we consider $(\tau \times f)^*(\gamma) \in H^n(B^n, \dot{B}^n; Z/2Z)$. The cap-product $((\tau \times f)^*(\gamma) \cap \sigma_n)$ lies in $H_0(B^n; Z/2Z)$ so by application of the augmentation homomorphism we can put

$$T_\gamma([B^n, f]_2) = \varepsilon_*((\tau \times f)^*(\gamma) \cap \sigma_n) \in Z/2Z.$$

It is not difficult to show that T_γ is a well defined element in $hom^n(X,A)$. Since

$$H^n(BO \times X,\ BO \times A;\ Z/2Z) \simeq \sum_{p+q=n} H^p(BO;Z/2Z) \otimes H^q(X,A;Z/2Z)$$

this is actually just a method for defining the mod 2 characteristic numbers of a map in the relative case. Specifically, if $\gamma = V_\omega \times h_q$, $|\omega| = p$, then $T_\gamma([B^n,f]_2) = \langle W_\omega f^*(h_q), \sigma_n\rangle \in Z/2Z$. Also, since (17.2) and (17.3) extend to the relative case, we find that

$$H^n(BO \times X,\ BO \times A;\ Z/2Z) \to hom^n(X,A)$$

is an epimorphism. There is, however, a non-trivial kernel $Wu^n(X,A) \subset H^n(BO \times X,\ BO \times A;\ Z/2Z)$ which we call the *generalized Wu relations*. If $X = \{pt\}$ then $Wu^n(pt)$ is the kernel of the epimorphism

$$H^n(BO;Z/2Z) \to Hom_{Z/2Z}(MO_n;Z/2Z).$$

Thus $Wu^n(pt)$ are just the Wu relations discussed by Dold in [Do_3]. Clearly $Wu^*(X,A)$ is itself a generalized cohomology theory, however, as we shall see in the following section, it does not admit an elementary description similar to $MO^*(X,A)$.

Now for each $\omega = (i_1 \le \dots \le i_k)$ containing no term of the form $2^j - 1$ (these are called by Thom the non-dyadic partitions) we consider the associated polynomial $s_\omega \in H^{|\omega|}(BO;Z/2Z)$ in the universal Whitney classes corresponding to the symmetric function $\sum t_1^{i_1} \dots t_k^{i_k}$. Let $S^n \subset H^n(BO;Z/2Z)$ be the subspace spanned by the s_ω for all non-dyadic partitions of n. Thom showed that under $H^n(BO;Z/2Z) \to hom^n(pt)$ the subspace S^n is carried isomorphically onto $Hom(MO_n,Z/2Z)$, [T_2]. In general let $S^n(X,A) \subset H^n(BO \times X,\ BO \times A;\ Z/2Z)$ be given by

$$S^n(X,A) = \sum_0^n S^k \otimes H^{n-k}(X,A;Z/2Z).$$

We claim that under the natural transformation

$$T : H^n(BO \times X,\ BO \times A;\ Z/2Z) \to hom^n(X,A)$$

the subspace $S^n(X,A)$ is always carried isomorphically onto $hom^n(X,A)$. From consideration of the dimensions of the $Z/2Z$ vector spaces $S^n(X,A)$ and $hom^n(X,A)$ it is only necessary to show that $Wu^n(X,A) \cap S^n(X,A) = \{o$ This follows easily by an argument analogous to that used in (17.2) and (17.3). Thus we have a splitting

$$H^n(BO \times X,\ BO \times A;\ Z/2Z) = Wu^n(X,A) \oplus S^n(X,A).$$

As we shall see this is a natural decomposition.

Consider now a map $\varphi : (X,A) \to (Y,B)$ which by duality induces $\varphi^* : \hom^n(Y,B) \to \hom^n(X,A)$. There is also the map

$$\Phi = id \times \varphi : (BO \times X,\ BO \times A) \to (BO \times Y,\ BO \times B)$$

which leads to a commutative diagram

$$\begin{array}{ccc} H^n(BO \times X,\ BO \times A;\ Z/2Z) & \longrightarrow & \hom^n(X,A) \\ \uparrow \Phi^* & & \uparrow \varphi^* \\ H^n(BO \times Y,\ BO \times B;\ Z/2Z) & \longrightarrow & \hom^n(Y,B). \end{array}$$

By duality, the kernel of φ^* consists exactly of these homomorphisms which annihilate the image of $\varphi^* : MO_n(X,A) \to MO_n(Y,B)$. We assert that under the natural transformation

$$T : H^n(BO \times Y,\ BO \times B;\ Z/2Z) \to \hom^n(Y,B)$$

the kernel of Φ^* is carried onto the kernel of φ^*. Suppose $c_n \in H^n(BO \times Y,\ BO \times B;\ Z/2Z)$ has $\varphi^* T(c_n) = o$, then write $c_n = a_n + b_n$ with $a_n \in S^n(Y,B)$ and $b_n \in Wu^n(Y,B)$. By hypothesis, $\Phi^*(c_n) \in Wu^n(X,A)$, but $\Phi^*(a_n) \in S^n(X,A)$ and $\Phi^*(b_n) \in Wu^n(X,A)$. Therefore $\Phi^*(a_n) = o$ and of course $T(c_n) = T(a_n) \in \hom^n(Y,B)$. As an immediate corollary we find

(17.4) THEOREM: *Let* $\varphi : (X,A) \to (Y,B)$ *be a map between finite* CW-pairs. *The necessary and sufficient condition that* $[B^n,f]_2 \in MO_n(Y,B)$ *lie in the image of* $\varphi^* : MO_n(X,A) \to MO_n(Y,B)$ *is that every characteristic number of* $[M^n,f]_2$ *associated with an element in the kernel of*

$$\varphi^* : H^*(Y,B;Z/2Z) \to H^*(X,A;Z/2Z)$$

must vanish. ∎

For the remainder of this section we can proceed to discuss the oriented bordism groups $MSO_n(X)$ by analogy. To each partition $\omega = (i_1 \le \dots \le i_k)$ we can associate $P_\omega = p_{i_1} \dots p_{i_k}$, the product of the Pontrjagin classes of a closed oriented manifold, in $H^{4|\omega|}(M^n;Z)$. Let $h \in H^m(X;Z)$ be an integral cohomology class with $m + 4|\omega| = n$. For an oriented singular manifold $f : M^n \to X$ we then have the numbers $\langle P_\omega f^*(h), \sigma_n\rangle \in Z$; we call these the Pontrjagin numbers of the map f.

Since $MSO_n(X) \simeq \sum_{p+q=n} H_p(X;MSO_q)$ mod C it follows that

$$MSO_n(X) \otimes Q \simeq \sum_{p+q=n} H_p(X;Q) \otimes (MSO_q).$$

Just as for (17.3) it now follows

(17.5) LEMMA: *Two oriented singular* n-*manifolds in* X *define the same element in* $MSO_n(X) \otimes Q$ *if and only if their corresponding Pontrjagin numbers are equal.* ∎

To close this section we shall use this to show in which cases the Whitney and Pontrjagin numbers of a map can be used to uniquely determine oriented bordism classes.

(17.6) THEOREM: *Let* X *be a finite* CW-*complex for which the torsion subgroup of* $H_*(X,Z)$ *is an elementary abelian* 2-*group. Then two oriented singular* n-*manifolds in* X *represent the same element in* $MSO_n(X)$ *if and only if they have the same Pontrjagin numbers and the same Whitney numbers.*

PROOF: In view of (15.2), $MSO_n(X) \simeq \sum_{p+q=n} H_p(X;MSO_q)$ and thus the torsion subgroup of $MSO_n(X)$ also is an elementary abelian 2-group. Suppose all the Whitney numbers and all the Pontrjagin numbers of $[M^n,f]$ vanish. Then by (17.3), $r[M^n,f] = [M^n,f]_2 = o$. Thus by the exact Rochlin sequence (16.2) we can write $[M^n,f] = 2[V^n,g]$. However, by (17.4) the class $[M^n,f]$ is a torsion class, as is $[V^n,g]$, so that $2[V^n,g] = o = [M^n,f]$. ∎

18. Generalized Wu relations

In the first edition of this book we were unable to describe the generalized Wu relations. It was Brown and Peterson who first characterized $Wu^*(X)$, [BP]. We shall include a brief discussion of their work both because Brown-Peterson draw significant corollaries from their determination of $Wu^*(X)$ and because this will afford us an opportunity to at least mention some of the fundamental relations between characteristic classes and cohomology operations.

In this section we denote by A the mod 2 Steenrod algebra with whose structure we shall require a minimal acquaintance, $[M_2]$. The canonical anti-automorphism of A is denoted by $\chi : A \to A$.

Let us return now to consideration of Thom spaces associated to orthogonal vector bundles. To each orthogonal k-plane bundle $\xi \to X$ there is, in the following manner, associated a corresponding right graded A-module structure on $H^*(X;Z/2Z)$. If $T(\xi)$ is the Thom space and $U \in \tilde{H}^k(T(\xi);Z/2Z)$ the mod 2 Thom class then $U \cup H^*(X;Z/2Z) \simeq H^*(T(\xi);Z/2Z)$. For any $x \in H^*(X;Z/2Z)$ and any operation $a \in A$ we define

$(x)a \in H^*(X;Z/2Z)$ to be the unique class which satisfies

$$U \cdot (x)a = \chi(a)(U \cdot x).$$

Three comments follow immediately from the definition

(i) *The Thom isomorphism* $H^*(X;Z/2Z) \simeq H^*(T(\xi);Z/2Z)$ *is an "anti-isomorphism" between the right graded A-module structure on* $H^*(X;Z/2Z)$ *and the left graded A-module structure on* $\tilde{H}^*(T(\xi);Z/2Z)$.

(ii) *If*

$$\begin{array}{ccc} \xi & \longrightarrow & \xi_1 \\ \downarrow & & \downarrow \\ X & \xrightarrow{f} & X_1 \end{array}$$

is a bundle map between k-plane bundles then the induced $f^* : H^*(X_1;Z/2Z) \to H^*(X;Z/2Z)$ *is a homomorphism of right A-module structures.*

(iii) *The induced right A-module structure on* $H^*(X;Z/2Z)$ *depends only on the stable equivalence class of* $\xi \to X$.

Recalling that Thom showed $H^*(MO;Z/2Z)$ is a free left A-module and combining the three preceding remarks we obtain

(18.1) LEMMA: *There is a free graded right A-module structure on* $H^*(BO;Z/2Z)$. ∎

It immediately follows that on the category of finite CW-pairs the functor $H^*(BO;Z/2Z)_A \otimes \tilde{H}^*(X/A;Z/2Z)$ is a generalized cohomology theory. The only point that might have been in doubt is exactness, but this is settled by (18.1). The object now is to show that there is a natural isomorphism

$$H^*(BO;Z/2Z)_A \otimes \tilde{H}^*(X/A;Z/2Z) \simeq \widetilde{hom}^*(X/A).$$

We observe that since X/A always has a base point we have a splitting $hom^*(X/A) = \widetilde{hom}^*(X/A) \oplus hom^*(pt)$. In addition, $\widetilde{hom}^n(X/A)$ is identified with $Hom_{Z/2Z}(\widetilde{MO}_n(X/A);Z/2Z)$.

Let us specialize X to a finite CW-complex and prove a computational lemma. On a finite CW-complex an automorphism of $H^*(X;Z/2Z)$ can be defined by $Sq(x) = \sum_0^\infty Sq^i(x)$. Similarly we could introduce $\chi(Sq)(x) = \sum_0^\infty \chi(Sq^j)(x)$. However, the defining properties of the canonical anti-automorphism show $Sq(\chi(Sq)(x)) = x$ so that in fact $Sq^{-1} = \chi(Sq)$. Thom proved that $S_q(U) = U \cup V$ where $V = (1, v_1, \ldots, v_k)$ is the total Whitney class of the bundle $[T_1]$ and accordingly we can state

(18.2) LEMMA: *Let* $\xi \to X$ *be a* k-*plane bundle over a finite* CW-*complex, then for any* x *in* $H^*(X;Z/2Z)$

$$(x)Sq = u \cdot Sq^{-1}(x)$$

where $u \in H^*(X;Z/2Z)$ *is the unique cohomology class for which* $S_q(u) = \bar{V}$, *the total dual Whitney class.*

PROOF: By definition $U \cdot ((x)Sq) = \chi(Sq)(U \cdot x)$ and since $\chi(Sq) = Sq^{-1}$ we find $U \cdot (xSq) = Sq^{-1}(U) \cdot Sq^{-1}(x)$. Clearly $U = Sq^{-1}Sq(U) = Sq^{-1}(U) \cdot Sq^{-1}(V)$ implies $Sq^{-1}(U) = U \cdot u$ where $Sq(u) = \bar{V}$, the dual Whitney class. ∎

Now suppose that $\eta \to M^n$ is the (stable) normal bundle to a closed n-manifold. In this case $\bar{V}$ will be the total tangential Stiefel-Whitney class of the manifold and u will be the total Wu class. Let us see this will imply.

First, there is the fundamental formula of Wu which relates Sq to the total Wu class u [Wu]. We state this in the following form.

(18.3) WU: *For any element* $x \in H^*(M;Z/2Z)$

$$\langle Sq(x), \sigma_n \rangle = \langle u \cdot x, \sigma_n \rangle$$

in Z/2Z. ∎

From this we can now derive the key lemma needed to characterize the generalized Wu relations.

(18.4) BROWN-PETERSON: *If the right action of* A *on* $H^*(M;Z/2Z)$ *is defined by the stable normal bundle, then for any operation* $a \in A$ *and any pair of elements* x,y *in* $H^*(M;Z/2Z)$

$$\langle (xa) \cdot y, \sigma_n \rangle = \langle x \cdot (ay), \sigma_n \rangle .$$

PROOF: Let us consider Sq first. By (18.2)

$$\langle (xSq) \cdot y, \sigma_n \rangle = \langle u \cdot (Sq^{-1}(x)) \cdot y, \sigma_n \rangle .$$

Upon application of the Wu formula (18.3) we can write

$$\langle u \cdot (Sq^{-1}x \cdot y), \sigma_n \rangle = \langle xSqy, \sigma_n \rangle .$$

Since this is an identity for all x,y in H*(M;Z/2Z) it will follow that $<(xSq^j)y,\sigma_n> = <x(Sq^j y),\sigma_n>$ for each $j \geq o$. Finally we appeal to the fact that A is generated as an algebra over Z/2Z by the operations Sq^j to complete the proof. ∎

Fix an integer $n \geq o$. For any finite CW-pair there is an epimorphism

$$H^*(BO;Z/2Z) \times \tilde{H}^*(X/A) \to \mathrm{Hom}(\tilde{MO}_n(X/A;Z/2Z).$$

This time we use the map $\eta : M^n \to BO$ which classifies the stable normal bundle so that $\eta^* : H^*(BO;Z/2Z) \to H^*(M;Z/2Z)$ is a homomorphism of right A-modules. Given $x \times y$ in the product $H^*(BO;Z/2Z) \otimes \tilde{H}^*(X/A;Z/2Z)$ and a singular manifold $f : M^n \to X/A$ for which $[M^n]_2 = o$ in MO_n we define

$$T_{x\times y}([M^n,f]_2) = <\eta^*(x)f^*(y),\sigma_n>.$$

It is important to note that for any operation $a \in A$, $<\eta^*(xa)f^*(y),\sigma_n> = <(\eta^*(x)a)f^*(y),\sigma_n> = <\eta^*(x)(af^*(y)),\sigma_n> = <\eta^*(x)f^*(ay),\sigma_n>$. Thus $xa \otimes y$ and $x \otimes ay$ define the same homomorphism. So that in fact for each $n \geq o$ we have an epimorphism

$$H^*(BO;Z/2Z)_A \otimes \tilde{H}^*(X/A;Z/2Z) \to \widetilde{\mathrm{hom}}^n(X/A).$$

We should also see that any homogeneous element of degree different from n lies in the kernel of this epimorphism, so that in fact there is really a degree preserving epimorphism

$$H^*(BO;Z/2Z)_A \otimes \tilde{H}^*(X/A;Z/2Z) \to \widetilde{\mathrm{hom}}^*(X/A).$$

Thom showed in $[T_2]$ that this is an isomorphism when X = {pt} and $A = \emptyset$; that is, we have an isomorphism of coefficient groups. We leave it to the reader to show in the fashion of (12.9) that this yields therefore a natural isomorphism of cohomology theories.

(18.5) BROWN-PETERSON: *The kernel of the epimorphism*

$$H^*(BO;Z/2Z) \otimes H^*(X;Z/2Z) \to H^*(BO;Z/2Z)_A \otimes H^*(X;Z/2Z).$$

is the generalized Wu relations. ∎

We can also put this in the form

(18.6) COROLLARY: *For* $n \geq o$

$$Wu^n(X) \subset \sum_{p+q=n} H^p(BO;Z/2Z) \otimes H^q(X;Z/2Z)$$

is the subspace space spanned by all elements $xSq^k \otimes y + x \otimes Sq^k y$ *where* $x \in H^i(BO;Z/2Z)$, $y \in H^j(X;Z/2Z)$ *and* $i + j + k = n$. ∎

19. The existence of an MSO_*-base

We saw in (8.3) and (17.3) that $MO_*(X)$ is always a free graded MO_*-module. This is not generally true for $MSO_*(X)$ but we can give a special case in which we find $MSO_*(X)$ is free over MSO_*.

(19.1) THEOREM: *If X is a finite CW-complex for which $H_*(X;Z)$ has no torsion, then $MSO_*(X)$ admits a homogeneous MSO_*-module basis and hence is a free graded MSO_*-module.*

PROOF: Let $\left\{c_{n,i}\right\}$ denote an additive homogeneous base for $H_*(X;Z)$. According to (15.2) there are oriented singular n-manifolds $f : M_i^n \to X$ with $\mu[M_i^n,f] = c_{n,i}$. We shall show that $\left\{[M_i^n,f]\right\}$ forms a homogeneous MSO_*-basis of $MSO_*(X)$.

We rely on the triviality of the bordism spectral sequence. There is the filtration $MSO_n(X) = J_{n,o} \supset \ldots \supset J_{o,n} \supset \{o\}$ with

$$J_{k,n-k}/J_{k-1,n-k+1} \simeq E_{k,n-k}^{\infty} \simeq E_{k,n-k}^2 \simeq H_k(X;MSO_{n-k}).$$

Since X has no torsion $H_k(X;MSO_{n-k})$ is isomorphic to $H_k(X;Z) \otimes MSO_{n-k}$ and thus $E_{k,o}^2 \otimes MSO_{n-k} \simeq E_{k,n-k}^2$.

Let $A \subset MSO_*(X)$ be the submodule generated by these $\left\{[M_i^n,f]\right\}$. We show by induction on k that $J_{k,n-k} \subset A$. Suppose this is true for $k - 1$. Choose $\alpha \in J_{k,n-k}$ so that $\hat{\alpha} \in E_{k,n-k}^2 \simeq J_{k,n-k}/J_{k-1,n-k+1}$ corresponds to α. Since $E_{k,o}^2 \otimes MSO_{n-k} \simeq E_{k,n-k}^2$, $\hat{\alpha}$ can be uniquely expressed as $\sum_i c_{k,i} \otimes [V_i^{n-k}]$. There is then $\beta = \sum_i [M_i^n,f][V_i^{n-k}]$ in $MSO_n(X)$. We may take $f : M_i^k \to X$ to be a map into the k-skeleton so that $\beta \in J_{j,n-k}$. But β also corresponds to $\hat{\alpha} \in E_{k,n-k}^2$ so that $\alpha - \beta \in J_{k-1,n-k+1}$. By induction then $A = MSO_*(X)$.

The independence can be seen as follows. If $\sum [M_i^{n-m},f][V_i^m] = o$ use (17.5) to show all $[V_i^m]$ are torsion classes of order 2. Since $\sum [M_i^{n-m}]_2 [V_i^m]_2 = o$ it will follow from (17.3) that $[V_i^m]_2 = o$ and hence $[V_i^m] = o$. ∎

At this point we shall add in a *Künneth formula* which will be used in Chapter III. In Section 6 we mentioned a homomorphism

$$\chi : MSO_p(X) \otimes MSO_q(Y) \to MSO_{p+q}(X \times Y)$$

which is

$$\chi([V^p,f] \otimes [M^q,g]) = [V^p \times M^q, f \times g].$$

For any three spaces, X, Y, Z it is easy to see that the following diagram is commutative.

$$\begin{array}{ccc}
 & MSO_{p+q}(X \times Y) \otimes MSO_r(Z) & \\
\nearrow & & \searrow \\
MSO_p(X) \otimes MSO_q(Y) \otimes MSO_r(Z) & & MSO_{p+q+r}(X \times Y \times Z) \\
\searrow & & \nearrow \\
 & MSO_p(X) \otimes MSO_{q+r}(Y \times Z) &
\end{array}$$

Letting Y be a single point, commutativity in the above diagram shows that

$$\chi : MSO_*(X) \otimes MSO_*(Z) \to MSO_*(X \times Z)$$

can be factored through $MSO_*(X) \otimes_{MSO_*} MSO_*(Z)$. For tensor products of graded modules over a graded ring we might refer to [C_2]. We thus arrive at a homomorphism

$$\chi : MSO_*(X) \otimes_{MSO_*} MSO_*(Z) \to MSO_*(X \times Z).$$

We can prove

(19.2) THEOREM: *Suppose* Y *is a finite* CW-*complex with* $MSO_*(Y)$ *a free graded* MSO_*-*module. For each* CW-*complex* X, *the homomorphism*

$$\chi : MSO_*(X) \otimes_{MSO_*} MSO_*(Y) \to MSO_*(X \times Y)$$

is an isomorphism.

PROOF: For X a point the result is immediate since $MSO_*(pt) \cong MSO_*$. We next show an isomorphism

$$\chi : MSO_*(I^n, S^{n-1}) \otimes_{MSO_*} MSO_*(Y) \to MSO_*(I^n \times Y, S^{n-1} \times Y).$$

Consideration of the triple $I^n \times Y$, $S^{n-1} \times Y$, $I^{n-1} \times Y$ reveals a boundary isomorphism

$$MSO_*(I^n \times Y, S^{n-1} \times Y) \cong MSO_*(S^{n-1} \times Y, I^{n-1} \times Y) \cong$$

$$MSO_*(I^{n-1} \times Y, S^{n-2} \times Y).$$

The isomorphism $MSO_*(I^n, S^{n-1}) \cong MSO_*(I^{n-1}, S^{n-2})$ when tensored with $MSO_*(Y)$ gives a commutative diagram

$$\begin{array}{ccc}
MSO_*(I^n,S^{n-1} \otimes_{MSO_*} MSO_*(Y) & \longrightarrow & MSO_*(I^n \times Y,S^{n-1} \times Y) \\
\downarrow \simeq & & \downarrow \simeq \\
MSO_*(I^{n-1},S^{n-2}) \otimes_{MSO_*} MSO_*(Y) & \longrightarrow & MSO_*(I^{n-1} \times Y,S^{n-2} \times Y).
\end{array}$$

Therefore the isomorphism

$$\chi : MSO_*(I^n,S^{n-1}) \otimes_{MSO_*} MSO_*(Y) \simeq MSO_*(I^n \times Y,S^{n-1} \times Y)$$

follows by induction on n.

Next assuming X is finite, we shall induct on the number of cells of X. Suppose the theorem is true if X has no more k-1 cells, and consider now an X with k cells. There exists a closed subcomplex X_1 of X contain ing all but one of the cells of X. From the exact triangle

$$\begin{array}{ccc}
MSO_*(X_1) & \longrightarrow & MSO_*(X) \\
\nwarrow & & \swarrow \\
& MSO_*(X,X_1) &
\end{array}$$

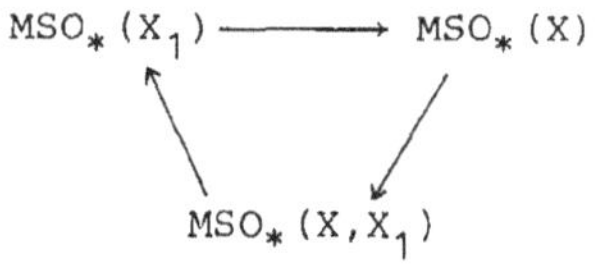

there results, since $MSO_*(Y)$ is free, an exact triangle

$$\begin{array}{ccc}
MSO_*(X_1) \otimes_{MSO_*} MSO_*(Y) & \longrightarrow & MSO_*(X) \otimes_{MSO_*} MSO_*(Y) \\
\nwarrow & & \swarrow \\
& MSO_*(X,X_1 \otimes_{MSO_*} MSO_*(Y). &
\end{array}$$

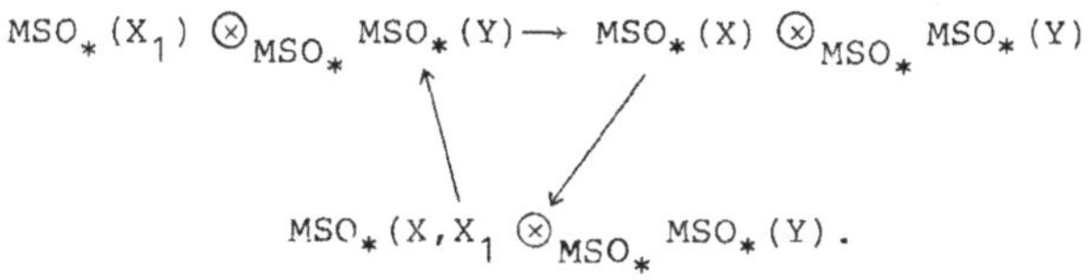

Understanding that tensor products are taken over MSO_* we have a commutative diagram

$$\begin{array}{ccccc}
MSO_*(X_1) \otimes MSO_*(Y) & & \longrightarrow & & MSO_*(X) \otimes MSO_*(Y) \\
\uparrow & \downarrow \chi_1 & & \swarrow \chi_2 & \downarrow \\
 & MSO_*(X_1 \times Y) & \rightarrow & MSO_*(X \times Y) & \\
 & \uparrow & & \swarrow & \\
 & & MSO_*(X \times Y,\ X_1 \times Y) & & \\
 & & \uparrow \chi_3 & & \\
\nwarrow & & MSO_*(X,X_1) \otimes MSO_*(Y) & & \swarrow
\end{array}$$

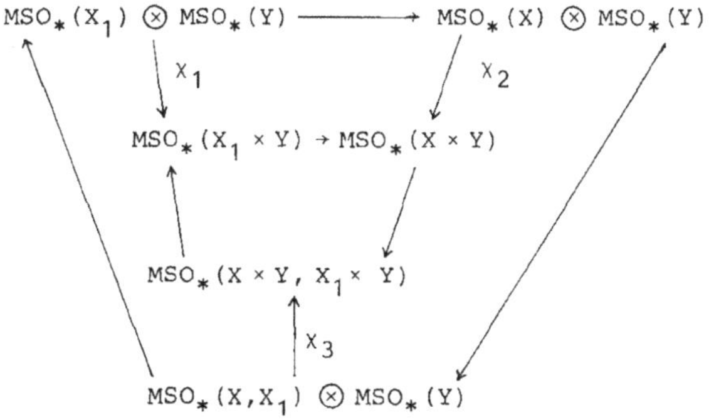

We have by induction that χ_1 is an isomorphism. Also (X,X_1) is relatively homeomorphic to (I^n,S^{n-1}) and $(X \times Y,\ X_1 \times Y)$ to $(I^n \times Y,\ S^{n-1} \times Y)$. Hence χ_3 is an isomorphism; therefore by the five-lemma χ_2 is also a isomorphism. Thus the theorem follows for X finite. The extension to an arbitrary complex is left to the reader.

In effect we have set up two generalized homology theories. The first is $MSO_*(X \times Y, A \times Y)$ and the second is $MSO_*(X,A) \otimes_{MSO_*} MSO_*(Y)$. The freeness of $MSO_*(Y)$ guarantees the exactness in the second case. Then χ becomes a natural transformation which is an isomorphism for a point.

(19.3) EXERCISE: Set up a Künneth formula for unoriented bordism. ∎

CHAPTER II. DIFFERENTIABLE INVOLUTIONS

After a preliminary section we shall settle down to a detailed study of diffeomorphisms of period 2 on closed unoriented manifolds. We shall simply call these *involutions*. The chief objective is to relate the fixed point set of an involution to the unoriented bordism class of the closed manifold on which that involution is defined. The first step, however, is the determination of the bordism module $MO_*(C_2)$ of all fixed point free involutions on closed unoriented manifolds. It is our common approach to periodic maps to first develop the fixed point free case and then to apply the resulting information to the study of fixed point sets.

Next we shall go into $I_*(C_2)$, the bordism algebra of all possible involutions on closed manifolds. The major effort of this chapter will be concentrated at this point with the objective of presenting Boardman's result [Ba].

An expanded collection of examples and special cases are included in this chapter. There is a report on the actions of elementary abelian 2-groups and the chapter is closed with an appendicial section discussing a generalization of the well known Borsuk-Ulam mapping theorem.

20. Preliminaries on group actions

Let G be a compact Lie group. Let B^n be a compact manifold. A *differentiable action* of G on B^n is a smooth map $\eta : G \times B^n \to B^n$ such that

i) $\eta(e,x) = x$, for $e \in G$ the identity.

ii) $\eta(g_1,\eta(g_2,x)) = \eta(g_1g_2,x)$.

Such *smooth actions* are basically the topic of this tract.

(20.1) LEMMA: *Suppose G acts smoothly on the compact n-manifold B^n and also on the closed manifold V^m. If A is a closed invariant subset of B^n and if $f : A \to V^m$ is smooth and equivariant then there is an open invariant set W, $A \subset W \subset B^n$ and an equivariant smooth extension $F : W \to V^m$.*

PROOF: By the term *equivariant* we mean $f(gx) = gf(x)$ for all $g \in G$ and all x in the domain of f.

First we may assume $B^n \subset M^n$, a closed n-manifold (for example if $\dot{B}^n \neq \emptyset$, then take M^n to be the double of B^n). Now at each point $(g,x) \in G \times B^n$ there is a neighborhood in $G \times M^n$ to which η may be smoothly extended. From Milnor [M_6] then there will be an open neighborhood U, $B^n \subset U \subset M^n$ and a smooth extension $\tilde{\eta} : G \times U \to M^n$ of η. We do not claim that $\tilde{\eta}$ is a group action.

Now to the proof of (20.1). Choose an open set $W_1 \subset M^n$ with $A \subset W_1 \subset U$ and a smooth extension $\tilde{f} : W_1 \to V^m$ of the map f. There exists an open set $W_2 \subset M^n$ with $A \subset W_2 \subset W_1 \subset U$ and $\tilde{\eta}(G \times W_2) \subset W_1$ because A is G-invariant.

Next we must appeal to the Mostow-Palais embedding theorem [Mo, P]. There is an orthogonal representation of G on R^k, for some $k > 0$, and an equivariant differentiable embedding $\varphi : V^m \to R^k$. There is also an open invariant set $S \supset \varphi(V^m)$ and a smooth equivariant retract $\rho : S \to \varphi(V^m)$.

Denote by $\tilde{\tilde{f}}(g,x)$ the composite

$$G \times W_2 \xrightarrow{\tilde{\eta}} W_1 \xrightarrow{\varphi\tilde{f}} \varphi(V^m) \subset R^k.$$

Define

$$\tilde{F}(x) = \int_G g^{-1}\tilde{\tilde{f}}(g,x)dg;$$

then $\tilde{F}$ is a smooth map of W_2 into R^k with $\tilde{F} = \varphi f$ on A. Choose an invariant set W, open in B^n, with $A \subset W \subset W_2$ and with $\tilde{F}(W) \subset S$. The required extension of $f : A \to V^m$ is $\varphi^{-1}\rho\tilde{F} : W \to V^m$. ∎

As an immediate corollary we now have the following

(20.2) LEMMA: *If* G *acts smoothly on the compact* n*-manifold* B^n *then there is an open invariant set* W, $\dot{B}^n \subset W \subset B^n$, *and an equivariant differentiable retraction* $r : W \to \dot{B}^n$. ∎

We can now prove the *Equivariant Collaring Theorem*. In the statement it will be understood that G acts on $\dot{B}^n \times [0,1)$ as $g(x,t) = (gx,t)$.

(20.3) THEOREM: *Suppose that* G *acts smoothly on the compact* n*-manifold* B^n. *There is an open invariant set* V, *with* $\dot{B}^n \subset V \subset B^n$, *and an equivariant diffeomorphism* $h : V \to \dot{B}^n \times [0,1)$ *with* $h(x) = (x,0)$ *for all* $x \in \dot{B}^n$

PROOF: Consider the tangent bundle $\tau \to B^n$, which we may regard as the restriction to B^n of the tangent bundle to M^n. The group G acts as a group of bundle maps on $\tau \to B^n$ covering the action of G on B^n. For a smooth real valued function f and for a tangent vector $v \in B^n_x$ (the tangent space at x), denote by $\langle f,v\rangle$ the directional derivative of f in the direction v. For $g \in G$ denote by fg the composite function f(gx). Then we have the relation $\langle fg,v\rangle = \langle f,gv\rangle$. Next let $F(x) = \int_G f(gx)dg$. Then $\langle F,v\rangle = \int_G \langle fg,v\rangle dg = \int_G \langle f,gv\rangle dg$.

Let V_1 be an open neighborhood of $\dot{B}^n$ in B^n for which there is a diffeomorphism $h : V_1 \to \dot{B}^n \times [o,1)$ (refer to (1.2)). Let $f : B^n \to R$ be a smooth map which on V_1 is a projection of h(x) into [o,1). At a point $x \in \dot{B}^n$ we would then have $\langle f,v\rangle > o$ if v is a tangent vector pointing into the interior of B^n. Moreover, $\langle f,v\rangle = o$ if and only if v is tangent to $\dot{B}^n$. Now introduce $F(x) = \int_G f(gx)dg$. Then for $x \in \dot{B}^n$ and v a tangent vector pointing into the interior of B^n

$$\langle F,v\rangle = \int_G \langle f,gv\rangle dg > o.$$

Moreover, $\langle F,v\rangle = o$ if and only if v is tangent to $\dot{B}^n$.

Select an open invariant W, $\dot{B}^n \subset W \subset V_1$, for which there is an equivariant smooth retraction $r : W \to \dot{B}^n$. Next define $h : W \to \dot{B}^n \times [o,\infty)$ by $h(x) = (r(x),F(x))$. Along $\dot{B}^n$, h induces an isomorphism of tangent spaces; for $x \in \dot{B}^n$, $h(x) = (x,o)$ and h is a diffeomorphism of $\dot{B}^n$ onto $\dot{B}^n \times o$. There is then an open invariant W_1, $\dot{B}^n \subset W_1 \subset W$, such that $h : W_1 \to \dot{B}^n \times [o,\infty)$ is an equivariant diffeomorphism onto an open subset. Choose $\varepsilon > o$ so that $\dot{B}^n \times [o,\varepsilon)$ lies in the image $h(W_1)$ and let $V = h^{-1}(\dot{B}^n \times [o,\varepsilon))$. Then finally $h' : V \to \dot{B}^n \times [o,1)$ is $h'(x) = (r(x), \varepsilon^{-1}F(x))$. The theorem follows. ∎

Using this we are able to define the unrestricted bordism group of smooth actions of G on closed manifolds. Consider first the oriented case. Then (G,M^n) is a smooth action of G on a closed oriented n-manifold so that for every $g \in G$ the diffeomorphism $x \to gx$ is orientation preserving. We say (G,M^n) bounds if and only if there is an action (G,B^{n+1}) on a compact oriented (n+1)-manifold as a group of orientation preserving diffeomorphisms for which the induced action $(G,\dot{B}^{n+1})$ is equivariantly diffeomorphic to (G,M^n) by an orientation preserving diffeomorphism. We take $-(G,M^n)$ to be $(G,-M^n)$ by just reversing the orientation of M^n. A disjoint union $(G,M_1^n \sqcup M_2^n)$ can be formed in the obvious manner. We say then that (G,M_1^n) is bordant to (G,M_2^n) if and only if the disjoint union $(G, M_1^n \sqcup -M_2^n)$ bords. In view of (20.3) this is shown to be an equivalence relation just as ordinary bordism was argued to be in Section 2. The resulting set of equivalence classes is denoted by $\mathcal{O}_n(G)$. By disjoint union this becomes an abelian group with $-[G,M^n] = [G,-M^n]$.

There is an augmentation $\varepsilon : \mathcal{O}_n(G) \to MSO_n$ which is given by $\varepsilon[G,M^n] = [M^n]$. We did not require the action to be effective, so we also have $MSO_n \to \mathcal{O}_n(G)$ by taking the action of G to be trivial. Clearly we have a direct sum decomposition $\mathcal{O}_n(G) \simeq MSO_n \oplus \tilde{\mathcal{O}}_n(G)$ where $[G,M^n] \in \mathcal{O}_n(G)$ if and only if $[M^n] = o$.

The direct sum $\mathcal{O}_*(G) = \sum_0^\infty \mathcal{O}_n(G)$ is a graded commutative algebra with identity over MSO_*. Given $[G,M^n]$ and $[G,V^m]$ we set $[G,M^n][G,V^m] = [G,M^n \times V^m]$ where the diagonal action $g(x,y) = (gx,gy)$ on the product is understood. There is the relation $[G,M^n][G,V^m] = (-1)^{nm}[G,V^m][G,M^n]$, and the multiplicative identity is $[G,pt] \in \mathcal{O}_o(G)$. The MSO_*-module structure arises from the previously noted embedding $MSO_* \to \mathcal{O}_*(G)$.

We shall refer to $\mathcal{O}_*(G)$ as the *unrestricted oriented* G-*bordism algebra*. For finite cyclic groups much more is known about the additive and the MSO_*-module structures of $\mathcal{O}_*(G)$ than was known at the time of the first edition. We shall explore this later, but we still regard the ring structure in $\mathcal{O}_*(G)$ as being quite difficult to deal with.

There is an unoriented analogue of course. The *unrestricted unoriented* G-*bordism algebra* is denoted by $I_*(G)$. Every non-zero element has additive order 2. We shall be especially concerned with $I_*(C_2)$ in this chapter.

As our nomenclature suggests, it is possible to impose various restrictions on the types of actions allowed to obtain still other G-bordism groups. As an important example we shall now discuss the *principal* G-*bordism groups* under the restriction that G be finite. In this case the action (G,B^{n+1}) is orientation preserving and free; that is, every isotropy subgroup is trivial. Under this restriction we may proceed to define a bordism group $MSO_n(G)$ of principal orientation preserving actions of G on closed oriented n-manifolds. The direct sum $MSO_*(G)$ is made into a right graded MSO_* algebra as follows. Given $[G,M^n]$ and $[V^m]$ let G act on $M^n \times V^m$ by $g(x,y) = (gx,y)$. This is still a principal action and $[G,M^n][V^m] = [G,M^n \times V^m]$ defines the module structure. It is not really productive to impose on $MSO_*(G)$ an algebra structure by analogy with $\mathcal{O}_*(G)$. When G is abelian there is a product on $MSO_*(G)$ arising from the H-space structure on $K(G,1)$, but we confess our ignorance about this point.

If $B(G) = K(G,1)$ is the classifying space of G then we can state

(20.4) THEOREM: *For any finite group there is a natural isomorphism of* MSO_**-modules*

$$MSO_*(G) \simeq MSO_*(B(G))$$

which preserves dimension.

In other words, to deal with $MSO_*(G)$ we can use the bordism techniques from Chapter I.

PROOF: Consider then a universal principal G-bundle $\nu : E(G) \to B(G)$. An element of $MSO_n(B(G))$ is represented by a map $f : V^n \times B(G)$ from a closed oriented manifold. Then in $V^n \times E(G)$ let M^n be defined as the set of all (x,y) with $f(x) = \nu(y)$. Now (G,M^n) is the principal action $g(x,y) = (x,gy)$. This is nothing but the principal G-bundle over V^n pulled back by f. The quotient map $q : M^n \to V^n$ is $q(x,y) = x$ and this is a local homeomorphism. A differential structure is imposed on M^n to make q into a local diffeomorphism, which preserves orientation locally. In this fashion M^n becomes a closed manifold, oriented so that the degree of q is equal to the group order, together with a principal action of G as a group of orientation preserving diffeomorphisms. The correspondence $[V^n,f] \to [G,M^n]$ is easily seen to be well defined.

Conversely given (G,M^n) there is an equivariant map $F : (G,M^n) \to (G,E(G))$ which is unique up to equivariant homotopy. With $V^n = M^n/G$ there is induced a homotopically unique map $f : V^n \to B(G)$. Give V^n the differential structure to make q a local diffeomorphism and orient V^n so that the degree of q equals the order of G. This defines $[V^n,f] \in MSO_n(B(G))$. If there were a principal $(G,\dot{B}^{n+1}) = (G,M^n)$ then F extends to $\tilde{F} : (G,B^{n+1}) \to (G,B(G))$ and $f : M^n/G \to B(G)$ extends to $\tilde{f} : B^{n+1}/G \to B(G)$. Furthermore, B^{n+1}/G is a compact oriented manifold with boundary $\dot{B}^{n+1}/G = M^n/G = V^n$. ∎

From Section 7 we now find

(20.5) COROLLARY: *For any finite group there is a spectral sequence* $\{E^r_{p,q}\}$ *with* $E^2_{p,q} = H_p(G;MSO_q)$ *and whose* E^∞*-term is associated with a filtration of* $MSO_*(G)$. ∎

It is understood that each MSO_q is to be regarded as a trivial G-module. If, for example, G is a 2-*group* then by (15.2) the spectral sequence collapses and $MSO_n(G) \simeq \sum_{p+q=n} H_p(G;MSO_q)$.

The unoriented analogue is denoted by $MO_*(G)$ and of course by (8.3), $MO_n(G) \simeq \sum_{p+q=n} H_q(G;Z/2Z) \otimes MO_q$ for any finite group.

Just for the record let us now observe

(20.6) THEOREM: *Let* (G,M^n) *be a finite group of order* k *acting as a principal group of orientation preserving diffeomorphisms on a closed oriented manifold. If* M^n/G *is oriented so that the quotient map* $M^n \to M^n/G$ *has degree* k *then* $k[M^n/G] = [M^n]$ *in* MSO_n.

PROOF: This depends upon the fact that an oriented bordism class is uniquely determined by its Pontrjagin numbers and its Whitney-Stiefel numbers. Now the local diffeomorphism $q : M^n \to M^n/G$ pulls the tangent bundle of M^n/G back into the tangent bundle of M^n. Denote characteristic classes on M^n/G by $\tilde{p}_j$ and $\tilde{w}_j$ and those on M^n by p_j and w_j then it follows by naturality that $q^*(\tilde{p}_j) = p_j$ and $q^*(\tilde{w}_j) = w_j$. Denoting the orientation classes of M^n and M^n/G by σ and $\tilde{\sigma}$ respectively we also have $q_*(\sigma) = k\tilde{\sigma}$. If $n \equiv 0 \pmod 4$ then to every partition ω of $n/4$ there are associated the Pontrjagin numbers $\langle \tilde{P}_\omega, \tilde{\sigma} \rangle$ and $\langle P_\omega, \sigma \rangle$. Since $k\tilde{\sigma} = q_*(\sigma)$ and $q^*(\tilde{P}_\omega) = P_\omega$ we find $k\langle \tilde{P}_\omega, \tilde{\sigma} \rangle = \langle P_\omega, \sigma \rangle$. A similar argument applies to Whitney-Stiefel numbers. ■

To close out this preliminary section we shall again take up tubular neighborhoods from Section 10 but this time with a group acting. We are primarily concerned with the structure of the set of stationary points, F, of a group action on a closed manifold (G,M^n). That is, $F = \{x | x \in M^n \text{ and } gx = x \text{ for all } g \in G\}$. It is well known [MZ] that each component of F is a smooth closed submanifold of M^n. For each k, $o \leq k \leq n$, we denote by $F^k \subset M^n$ the union of the k-dimensional components of F, which are finite in number. Thus each F^k is a closed smooth k-dimensional submanifold (possibly disconnected, possibly empty).

Let us proceed now in a general vein. Let (G,M^n) be a closed G-manifold. By the usual averaging process there is a Riemannian metric on M^n with respect to which G acts as a group of isometries. Let $V^m \subset M^n$ be a closed regular submanifold invariant under the action of G. As in Section 10, there is the tubular neighborhood N of V^m with radius ε ($\varepsilon > o$ sufficiently small). Now N is always a regularly embedded compact n-dimensional submanifold. Since V^m is G-invariant and G is a group of isometries it is immediately clear that N is also invariant. If $V_1,\dots,V_k$ is a pairwise disjoint collection of closed invariant submanifolds of (G,M^n), possibly with varying dimension, then a tubular neighborhood of $V = \sqcup V_i$ is by definition the union of a pairwise disjoint collection of tubular neighborhoods of the various V_i.

Let us summarize the picture. Let $D(\eta) \to V^m$ be the normal (n-m)-disk bundle. Then G acts as a group of bundle maps on $D(\eta) \to V^m$ covering the action of G on the invariant V^m. Furthermore there is an equivariant diffeomorphism, which is the identity along V^m, with a compact invariant normal tube (G,N).

If $x \in M^n$ is a stationary point, then x can be considered as a G-invariant o-manifold and the preceding remarks applied. In this case the normal bundle is M^n_x, the space of tangent vectors to M^n at x and we receive an orthogonal representation of G on M^n_x. By the diffeomorphism h we map the unit disk of M^n_x, via an equivariant diffeomorphism onto an invariant closed neighborhood of x in M^n. This yields the local linearization of the action of G at a stationary point. From this it will immediately follow that the component of F containing x is a closed regular submanifold of M^n. Moreover, the tangent space to F at x is the subspace of M^n_x consisting of all the fixed vectors under the representation of G on M^n_x. The orthogonal complement of the fixed vectors in M^n_x is then the subspace of vectors at x orthogonal to the component of F containing x.

We wrote $F = \sqcup_0^n F^k$ and $\eta^{n-k} \to F^k$ will denote the normal bundle to the union of the k-dimensional components of the set of stationary points and $\eta \to F = \sqcup_0^n \eta^{n-k} \to F^k$ is just the normal bundle to the whole set of stationary points. The group G acts as a group of bundle maps on $\eta \to F$ sending each fibre onto itself. The representation in a fibre has no non-trivial fixed vector. We can identify $(G,D(\eta))$ with a closed invariant normal tube (G,N) around F. A particularly simple example occurs when $G = C_2$. Then C_2 acts orthogonally on the fibres of $\eta \to F$ leaving only the o-vector fixed. The only such representation of C_2 on a vector space is $v \to -v$.

21. Fixed point free involutions

We begin with a further analysis of $MO_*(C_2)$. Recall that $K(C_2,1) = B(C_2) = RP(\infty)$ and that $H^*(RP(\infty);Z/2Z) \simeq Z/2Z\,[c]$ is the polynomial ring on $c \in H^1(RP(\infty);Z/2Z)$. Now a fixed point free involution (T,X) is induced by a homotopically unique map $f : X/T \to RP(\infty)$. We shall write $c = f^*(c) \in H^1(X/T;Z/2Z)$ and call c the *characteristic class of the involution*. It is simply the Whitney class of the associated line bundle $\xi \to X/T$. Combining (20.3) and (17.3) we find

(21.1) THEOREM: *Let* (T,M^n) *be a fixed point free involution on a closed* n-*manifold. Then* $[T,M^n]_2 = o \in MO_n(C_2)$ *if and only if every Whitney number on* M^n/T *of the form* $\langle W_\omega c^m, \sigma(M/T)\rangle$ *vanishes, where* $\omega = (i_1 \le \ldots \le i_k)$ *is a partition of* $n-m$, $o \le m \le n$.

In general the $\langle W_\omega c^m, \sigma(M/T)\rangle \in Z/2Z$ will be called the *involution numbers* of (T,M^n) and they determine $[T,M^n]_2$ uniquely. It follows from (8.3) that $MO_*(C_2)$ is a free MO_*-module with one generator in each dimension.

(21.2) THEOREM: *Suppose for each non-negative integer* n *there is a fixed point free involution on a closed* n-*manifold such that for each* n *the involution number* $\langle c^n, \sigma(X^n/T)\rangle$ *of* (T,X^n) *is non-zero. Then* $\{[T,X^n]_2\}$ *is a homogeneous* MO_*-*module basis for* $MO_*(C_2)$. *In particular if* (A,S^n) *is the antipodal involution on the* n-*space then* $\{[A,S^n]_2\}$ *is a basis for* $MO_*(C_2)$.

PROOF: Consider $f: X^n/T \to RP(\infty)$ and note that $\langle c^n, f_*(\sigma(X^n/T))\rangle = \langle c^n, \sigma(X^n/T)\rangle \ne o$. The result now follows from (8.3). ∎

The reader is of course aware of the fact that on $RP(n) = S^n/A$, $c^n \ne o$. According to (21.2) for every fixed point free involution (T,M^n) there is a unique expression

$$[T,M^n]_2 = \sum_o^n [A,S^{n-m}]_2 [V^m]_2.$$

A most important example for us will be the *bundle involutions*. Consider a smooth orthogonal k-plane bundle $\xi \to V^n$ over a closed n-manifold. There is the associated sphere bundle $S(\xi) \to V^n$ with fibre S^{k-1}. The total space $S(\xi)$ is a closed $(n+k-1)$-dimensional manifold. The antipodal map $A: S^{k-1} \to S^{k-1}$ lies in the center of $O(k)$, and hence there is a fibre preserving fixed point free involution $(T,S(\xi))$ which on each fibre agrees with the antipodal map. We refer to $(T,S(\xi))$ as the bundle involution [CF_1].

Consider then the diagram

$$\begin{array}{ccc} S(\xi) & \longrightarrow & S(\xi)/T \\ & \searrow & \swarrow p \\ & V^n & \end{array}$$

We see that $p: S(\xi)/T \to V^n$ is a bundle with the real projective space $RP(k-1)$ as fibre and $PO(k)$ as structure group. We shall denote $S(\xi)/T \to V^n$ by $RP(\xi) \to V^n$; that is the *projective space bundle* associated to vector bundle. If $RP(k-1) \subset RP(\xi)$ is a fibre, then by naturality of

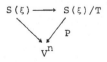

Whitney classes the image of $c \in H^1(RP(\xi);Z/2Z)$ under $i^* : H^1(RP(\xi);Z/2Z) \to H^1(RP(k-1);Z/2Z)$ is the generator of $H^*(RP(k-1);Z/2Z)$. The fibre is therefore totally non-homologous to zero. Via $p^* : H^*(V^n;Z/2Z) \to H^*(RP(\xi);Z/2Z)$ we may think of $H^*(RP(\xi);Z/2Z)$ as a graded module over the ring $H^*(V^n;Z/2Z)$. According to the Leray-Hirsch Theorem [B_2, p. 129], $H^*(RP(\xi);Z/2Z)$ is a free graded $H^*(V^n;Z/2Z)$-module with basis $1,c\dots,c^{k-1}$. Consider now the tangent bundle $\tau \to RP(\xi)$. The fibre map $p : RP(\xi) \to V^n$ splits τ into a Whitney sum $\tau = \tau_1 \oplus \tau_2$ [BH, p. 482]. Here τ_1 is the bundle pulled back by p from the tangent bundle to V^n. Put otherwise, τ_1 is the bundle of tangent vectors orthogonal to the fibre. The bundle $\tau_2 \to RP(\xi)$ is the bundle of tangent vectors parallel to the fibre. By naturality, the total Whitney class of τ_1 is $1 + p^*(w_1) + \dots + p^*(w_n)$ where the w_j are the tangential Whitney-Stiefel classes of V^n. The total Whitney class of τ_2 was calculated by Borel-Hirzebruch in [BH, p. 517].

(21.3) BOREL-HIRZEBRUCH: *The total Whitney class of the bundle* $\tau_2 \to RP(\xi)$ *is given by* $(1+c)^k + (1+c)^{k-1}p^*(v_1) + \dots + p^*(v_k)$, *where the* v_i *are the Whitney classes of* $\xi \to V^n$. Since τ_2 *is a* $(k-1)$*-plane bundle it also follows that*

$$c^k + p^*(v_1)c^{k-1} + \dots + p^*(v_k) = 0. \quad \blacksquare$$

The Stiefel-Whitney classes of the tangent bundle to $RP(\xi)$ are therefore given by

(21.4)

$$W_m = \sum_{p+q+r=m} \binom{k-p}{q} p^*(w_r v_p) c^q.$$

Actually this is not very useful in general. A better approach is the consideration of the factoring of a total Whitney class into a product of linear terms. Thus suppose

$$1 + w_1 + \dots + W_n = (1+t_1) \dots (1+t_n)$$

$$1 + v_1 + \dots + v_k = (1+\lambda_1) \dots (1+\lambda_k).$$

(21.5) COROLLARY: *In factored form the total Stiefel-Whitney class of the tangent bundle to* $RP(\xi)$ *may be expressed as*

$$(1+p^*(t_1)) \dots (1+p^*(t_n))(1+(c+p^*(\lambda_1))) \dots (1+(c+p^*(\lambda_k))). \quad \blacksquare$$

This is an immediate consequence of the Borel-Hirzebruch formula. We shall use (21.5) to construct some examples of closed manifolds with indecomposable bordism class. A bordism class in MO_n is *decomposable* if and only if it can be expressed as a sum of products of lower dimensional bordism classes; if this is not possible then the element of MO_n is said to be *indecomposable*. The importance of indecomposable classes lies in the fact that they may be used as generators of MO_* as a graded polynomial algebra over $Z/2Z$. We had already pointed out that $[RP(2k)]_2$ is indecomposable. Indecomposability is recognized as follows. Given M^n write the total Whitney-Stiefel class of the tangent bundle in factored form $(1+t_1)\ldots(1+t_n)$. Then $s_n \in H^n(M^n;Z/2Z)$ is the polynomial in the Whitney-Stiefel classes corresponding to the symmetric function $t_1^n + \ldots + t_n^n$. It was shown by Thom $[T_2]$ that $[M^n]_2$ is indecomposable if and only if $\langle s_n, \sigma(M^n)\rangle \neq o$ in $Z/2Z$.

For $r \geq o$ let $\xi \to RP(r)$ denote the real line bundle associated with the principal C_2-bundle $S^r \to S^r/A = RP(r)$. For $s \geq o$ consider the Whitney sum $\xi \oplus sR \to RP(r)$, where $sR \to RP(r)$ is the trivial s-plane bundle. We shall denote $RP(\xi \oplus sR)$ by $RP(r,s)$ which is a closed $(r+s)$-dimensional manifold.

(21.6) LEMMA: *If* $s > o$ *then* $[RP(r,s)]_2$ *is indecomposable if and only if* $s + \sum_0^r \binom{r+s}{j} \equiv 1 \pmod 2$.

PROOF: The total Whitney class of $\xi \oplus sR \to RP(r)$ is $1 + d$ where $d \in H^1(RP(r);Z/2Z)$ is the generator. Thus $\lambda_1 = d$, $\lambda_2 = \ldots = \lambda_{s+1} = o$. The tangential Stiefel-Whitney class of $RP(r)$ is $(1+d)^{r+1}$. Hence the total Stiefel-Whitney class of the tangent bundle to $RP(r,s)$ is

$$(1 + p^*(d))^{r+1}(1 + c)^s(1 + c + p^*(d)).$$

Therefore

$$s_{r+s} = (r+1)p^*(d^{r+s}) + sc^{r+s} + (c + p^*(d))^{r+s}.$$

Since $s > o, p^*(d^{r+s}) = o$. Referring to (21.3) we find $c^{s+1} = p^*(d)c^s$. From this we write

$$s_{r+s} = sp^*(d^r)c^s + \sum_0^r \binom{r+s}{j} p^*(d^r)c^s$$

$$= (s + \sum_0^r \binom{r+s}{j})p^*(d^r)c^s.$$

The assertion follows since the value of $p^*(d^r)c^s$ on the fundamental class of $RP(r,s)$ is 1 in $Z/2Z$. ∎ It may be noted that technically it is $\tau_1 \oplus R \to RP(r)$ which has class $(1+d)^{r+1}$ so that in fact we computed s_{r+s} for the bundle $(\tau_1 \oplus R) \oplus \tau_2 \to RP(r,s)$. This has no effect on the outcome.

(21.7) LEMMA: *If* $r = 2^k$ *and* $s = 2^{k+1}n - 1$ *with* $n > o$, $k > o$ *then* $[RP(2^k, 2^{k+1}n - 1)]_2$ *is indecomposable. If* $r = 2^k$, $s = 2^k(2n-1)$ *with* $n > o$, $k > o$, *then* $[RP(2^k, 2^k(2n-1))]_2$ *is indecomposable.*

PROOF: The binomial coefficient $\binom{r+s}{j} \equiv 1 \pmod 2$ if and only if every power of 2 occuring in the dyadic expression of j also occurs in the dyadic expression of $r+s$. Dyadic expression means to write the integer as a sum of powers of 2. If $r+s = 2^{k+1}n + 2^k - 1$ then $\binom{r+s}{j} \equiv 1 \pmod 2$ for $o \le j < 2^j = r$, but $\binom{r+s}{2^k} \equiv o \pmod 2$. So in this case $\sum_o^r \binom{r+s}{j} \equiv o$ (mod 2) while s is odd. Incidentally, this will furnish a list of odd dimensional generators for MO_*. In the next case $r+s = 2^{k+1}n$ so that $\binom{r+s}{j} \equiv o \pmod 2$ for $o < j \le 2k = r$ while $\binom{r+s}{o} = 1$. Of course s is even. ∎

These indecomposable bordism classes will reappear later in our discussion of Boardman's work. We end the section by posing an exercise for the reader. It can be done by computing characteristic numbers or it may be seen geometrically.

(21.8) EXERCISE: *Let* $\xi \to V^n$ *be a smooth* k*-plane bundle over a closed manifold. If either* $k = 2$ *or* $n = 1$ *then* $[RP(\xi)]_2 = o$. ∎

22. Fixed point sets of involutions

In this section we shall show just how the normal bundle to the fixed point set of an involution (T,M^n) on a closed manifold explicitly determines the mod 2 bordism class $[M^n]_2$. This will allow us to make use of any convenient involution to analyse the bordism class of a manifold. To illustrate the point we shall give a proof of the theorem of Wall that if M^n is a closed manifold then $M^n \times M^n$ is mod 2 bordant to an orientable manifold. A second illustration will be the study of fixed sets for conjugations on almost complex manifolds.

Consider a differentiable involution (T,M^n) on a closed manifold. We assume T is an isometry in a fixed Riemannian metric. Denote by F^k, $o < k \le n$, the union of the k-dimensional components in F. As in

Section 20, F^k is a closed regular submanifold. Note too that F^n consists of those components of M^n (if any) left pointwise fixed by T.

As mentioned also in Section 20, there are normal bundles $\eta^{n-k} \to F^k$ and $\eta \to F = \sqcup_0^n (\eta^{n-k} \to F^k)$, which is the normal bundle to all of F. We allow a o-plane bundle when n = k. There are also the bundle involutions $(T,S(\eta^{n-k}))$. For k < n the total space here has dimension n - 1. We set $(T,S(\eta)) = \sqcup_0^n (T,S(\eta^{n-k}))$. We call this $S(\eta)$ the *normal sphere bundle* to the fixed set. There is no contribution for n = k.

(22.1) LEMMA: *If* (T,M^n) *is an involution on a closed manifold with* $(T,S(\eta))$ *the bundle involution on the normal sphere bundle of the fixed set, then in* $MO_{n-1}(C_2)$

$$[T,S(\eta)]_2 = \textstyle\sum_0^{n-1} [T,S(\eta^{n-k})]_2 = o.$$

PROOF: We may as well assume $F^n = \emptyset$. Let (T,N) be a closed normal tube around F (Section 20). Then $W^n = M^n \setminus \mathrm{Int}(N)$ is a compact regular T-invariant n-dimensional submanifold of M^n for which (T,W^n) has no fixed points. The result now follows because

$$(T,\dot{W}^n) = (T,S(\eta)) = (T,\dot{N}). \quad ■$$

We come now to the chief result of this section.

(22.2) THEOREM: *Let* (T,M^n) *denote a smooth involution on a closed manifold with* $\eta \to F$ *the normal bundle to the fixed set. If* $\eta \oplus R \to F$ *is the Whitney sum of that normal bundle with a trivial line bundle then* $RP(\eta \oplus R)$ *is a closed* n*-manifold and* $[M^n]_2 = [RP(\eta + R)]_2$ *in* MO_n.

PROOF: The proof is obtained by considering the involutions $(T_1,M^n \times I)$ and $(T_2,M^n \times I)$ where $T_1(x,t) = (x,1-t)$, $T_2(x,t) = (t(x),1-t)$. The fixed set of T_1 is $M^n \times 1/2$ and the normal bundle to this fixed set is a trivial line bundle. The fixed set of T_2 is $F \times 1/2$. Identifying F with $F \times 1/2$, the normal bundle to the fixed set of T_2 is $\eta \oplus R \to F$.

Define an equivariant diffeomorphism $\varphi : (T_1, M^n \times \dot{I}) \to (T_2, M^n \times \dot{I})$ by $\varphi(x,1) = (Tx,1)$ and $\varphi(x,o) = (x,o)$. We adjoin $(T_1, M^n \times I)$ to $(T_2, M^n \times I)$ along their boundaries by φ. There results an involution (T_3, M^{n+1}) on a closed manifold. Now by application of (22.1) we find

$$[A,S^o]_2 \, [M^n]_2 + [T,S(\eta \oplus R)]_2 = o \in MO_n(C_2).$$

Passing to quotients therefore

$$[M^n]_2 = [RP(\eta \oplus R)]_2$$

in MO_n as required. ∎

This result was suggested by the easy observation that if a closed manifold admits a fixed point free involution then the manifold bounds mod 2. If (T,M^n) is the fixed point free involution then the mapping cylinder of the quotient map $M^n \to M^n/T$ is a compact (n+1)-manifold whose boundary is M^n. Thus (22.2) is the generalization of this to the case where fixed points are present.

(22.3) WALL: *If* M^n *is a closed manifold then* $M^n \times M^n$ *is unoriented bordant to an orientable manifold.*

PROOF: Consider on $M^n \times M^n$ the involution $T(x,y) = (y,x)$. The fixed point set is the diagonal Δ and the normal bundle to Δ in the product is equivalent to the tangent bundle $\tau \to M^n$. Thus by (22.2) we can say $[M^n]_2^2 = [RP(\tau \oplus R)]_2$. We can compute W_1 for $RP(\tau \oplus T)$ from (21.4) taking $k = n+1$. The answer is

$$W_1 = p^*(w_1) + p^*(w_1) + \binom{n+1}{1}c = \binom{n+1}{1}c.$$

Therefore if n is odd we simply conclude that $RP(\tau \oplus R)$ is orientable.

For n even we must be a bit more devious. On the complex projective space consider the conjugation involution $(T,CP(n))$ given by $T[z_1,\ldots,z_{n+1}] = [\bar{z}_1,\ldots,\bar{z}_{n+1}]$. The fixed point set is precisely the real projective $RP(n) \subset CP(n)$. We shall prove that the normal bundle to $RP(n)$ in $CP(n)$ is equivalent to the tangent bundle $\tau \to RP(n)$. At a point in $RP(n)$ consider the complex tangent space to $CP(n)$. The real vectors are identified with the tangent space to $RP(n)$ at this point, and the purely imaginary vectors make up the normal space. Then multiplication by $\sqrt{-1}$ will interchange the normal and tangential vectors to provide the required (real) equivalence. Now from (22.2) we find

$$[CP(n)]_2 = [RP(\tau \oplus R)]_2 = [RP(n)]_2^2.$$

This fact was known before by comparing characteristic numbers [Wa] and in fact Stong has provided an explicit bordism between $CP(n)$ and $RP(n)^2$ [St_6].

We can complete the argument now. Squaring in MO_* is a Frobenius endomorphism which doubles the degree of a homogenous element. Since MO_*

is a polynomial ring over $Z/2Z$ it is enough to check (22.3) on a set of generators. We have already disposed of the odd dimensions. For even dimensional generators we can use $[RP(2k)]_2$ and we just saw that $[CP(2k)]_2 = [RP(2k)]_2^2$. ∎

The introduction of the conjugation involution $(T, CP(n))$ suggests a generalization. Suppose $V^m \subset CP(n)$ is a non-singular projective variety embedded so that $T(V^m) = V^m$. Then the fixed set $F \subset V^m$ is called the real form of V^m. If $F \neq \emptyset$ then it is a closed submanifold of real dimension m and $[V^m]_2 = [F]_2^2$. Of course if $F = \emptyset$ then $[V^m]_2 = 0$ anyway.

We can put this matter into a more general setting. An almost complex structure on a manifold is a pair (M^{2m}, J) where J is a real linear bundle map of the tangent bundle

$$\begin{array}{ccc} \tau & \xrightarrow{J} & \tau \\ & \searrow \quad \swarrow & \\ & M^{2n} & \end{array}$$

which satisfies $J^2 = -I$, [S]. A differentiable involution T on M^{2n} induces a bundle map dT for which

$$\begin{array}{ccc} \tau & \xrightarrow{dT} & \tau \\ \downarrow & & \downarrow \\ M^{2m} & \xrightarrow{T} & M^{2n} \end{array}$$

We say T is a *conjugation of* (M^{2n}, J) if and only if $J \circ dT = -dT \circ J$.

(22.4) THEOREM: *If* (T, M^{2n}) *is a conjugation of the almost complex structure* (M^{2n}, J) *on a closed manifold and if* F *is the fixed set of* T, *then* F *is a closed* n-*manifold and* $[M^{2n}]_2 = [F]_2^2$ *in* MO_{2n}.

PROOF: This is valid if F is empty for in that case $[M^{2n}]_2 = 0$. Suppose then $F \neq \emptyset$ and take $x \in F$. In the tangent space M_x^{2n} introduce subspaces

$$F_x = \{v \mid dT(v) = v\} \qquad N_x = \{v \mid dT(v) = -v\}$$

This is a direct sum splitting of M_x^{2n} since for any $v \in M_x^{2n}$

$$v = ((v + dT(v))/2) + ((v - dT(v))/2).$$

Clearly F_x is the subspace of vectors tangent to F at x while N_x is the orthogonal vectors to F at x. Since $J \circ dT = -dT \circ J$ we find $J(F_x) = N_x$ and $J(N_x) = F_x$. Surely $\dim F_x = n$ and so F is an n-dimensional submanifold.

Since J provides a real equivalence between the tangential and normal bundles to F we may use (22.2) just as we did in (22.3) to see $[M^{2n}]_2 = [F]_2^2$. ∎

23. Normal and tangential bundles to fixed sets

In this section we make our first effort toward the problem of determining for an involution on a manifold (T,M^n) with $[M^n]_2 \neq o$ just how complicated the fixed set and its normal bundle must be. In this section we must introduce *bordism classes of vector bundles*.

Consider $MO_*(BO(k))$. An element of $MO_n(BO(k))$ is defined by a map $f : V^n \to BO(k)$. If two maps of V^n into $BO(k)$ are homotopic then they define the same bordism class. Thus we may think of a bordism class as given by a closed manifold, V^n, together with a homotopy class of maps into $BO(k)$. But the homotopy classes of maps into $BO(k)$ are in 1-1 correspondence with the orthogonal k-plane bundles over V^n. Therefore we receive a bundle interpretation for elements in $MO_n(BO(k))$. The bordism class is represented by an orthogonal k-plane bundle over a closed manifold; $\xi \to V^n$. Denote the bordism class of $\xi \to V^n$ by $[\xi \to V^n]_2$ or just $[\xi]_2$. A bundle bords if and only if there is an orthogonal k-plane bundle $\xi' \to B^{n+1}$ over a compact manifold with $\dot{B}^{n+1} = V^n$ and the restriction of ξ' to $\dot{B}^{n+1}$ is equivalent to $\xi \to V^n$. It follows that if ξ_1, ξ_2 are two k-plane bundles over V^n with the same Whitney numbers, then they are bordant as bundles. We shall agree that $MO_n(B(o)) = MO_n$.

If we wish to use smooth bundles in the above, we may replace $BO(k)$ by the Grassman manifold $M_{k,N}$ of unoriented k-planes through the origin in R^{k+N}. We take $N > n$ and use the unoriented differentiable bordism groups as defined in Section 9. In this manner we identify $MO_*(BO(k))$ with the bordism classes of smooth k-plane bundles over closed manifolds.

To each smooth orthogonal k-plane bundle we have the associated fixed point free bundle involution $(T,S(\xi))$. The assignment $[\xi]_2 \to [T,S(\xi)]_2$ yields a well defined homomorphism $\partial : MO_n(BO(k)) \to MO_{n+k-1}(C_2)$. In fact it is easily verified that $\partial : MO_*(BO(k)) \to MO_*(C_2)$ is an MO_*-homomorphism of degree k-1. It is agreed that this homomorphism is trivial if $k = o$. The homomorphism will be seen to be of central importance in the study of fixed point sets.

(23.1) THEOREM: *Suppose* (T,M^n) *is an involution on a closed manifold. Suppose that for each* m, $o \le m < n$, *all the normal Whitney classes of* $\eta^{n-m} \to F^m$ *vanish in positive dimensions. Then* $[F^m]_2 = o$ *for* $o \le m < n$ *and* $[M^n]_2 = [F^n]_2$.

PROOF: For $o \le m < n$ it follows from the hypothesis that the normal bundle $\eta^{n-m} \to F^m$ is bordant to the trivial (n-m)-plane bundle $(n-m)R \to F^m$. That is, $[\eta^{n-m}]_2 = [(n-m)R]_2[F^m]_2$. Therefore $\partial[\eta^{n-m}]_2 = [A,S^{n-m-1}]_2[F^m]_2$. Hence by (22.1)

$$o = [T,S(\eta)]_2 = \sum_o^{n-1}[A,S^{n-m-1}]_2[F^m]_2.$$

But the antipodal maps form a basis for $MO_*(C_2)$ and therefore $[F^m]_2 = o$ for $o \le m < n$. Furthermore, $[RP(\eta \oplus R)]_2 = [F^n]_2 + \sum_o^{n-1}[RP(n-m-1)]_2[F^m]_2 = [F^n]_2 = [M^n]_2$. ■

This theorem generalizes the observation that if (T,M^n), $n > o$, has only a finite number of fixed points then the number of fixed points is even and $[M^n]_2 = o$.

There is a sort of dual to (23.1) involving the vanishing of the Stiefel-Whitney classes of the tangent bundle of the fixed set. There is the important restriction that all components of the fixed point set have the same dimension, however.

(23.2) LEMMA: *Suppose* $\xi \to V^m$ *is a* k*-plane bundle over a closed, connected manifold. If all the positive dimensional Stiefel-Whitney classes of the tangent bundle to the base manifold vanish and if* $\partial[\xi]_2 = o \in MO_{n+k-1}(C_2)$ *then* $[\xi]_2 = o \in MO_m(BO(k))$.

PROOF: There is the bundle involution $[T,S(\xi)]_2 = o$ and $S(\xi)/T = RP(\xi)$. Every involution number of $[T,S(\xi)]_2$ vanishes and we shall prove that for any partition ω with $|\omega| = r$ the product $p^*(v_{i_1} \dots v_{i_k})c^{m+k-1-r} = o \in H^{m+k-1}(RP(\xi);Z/2Z)$. For $r = o$, $c^{m+k-1} = o$ since $\langle c^{m+k-1}, \sigma(RP(\xi))\rangle = o$ and $RP(\xi)$ is connected. Suppose by way of induction the assertion has been shown for $r < r_o$. Choose a partition ω with $|\omega| = r_o$, then $W_{i_1} \dots W_{i_k} c^{m+k-1-r_o} = o$, where W_i denotes the Stiefel-Whitney classes of the tangent bundle to $RP(\xi)$, because this is an involution number. Applying (21.4) in this case we find $W_i = p^*(v_i)$ + terms involving positive powers of c. Now then $W_{i_1} \dots W_{i_k} c^{m+k-1-r_o} = p^*(v_{i_1} \dots v_{i_k})c^{m+k-1-r_o}$ + terms involving a higher power of c. Inductively the remaining terms are eliminated so that $p^*(v_{i_1} \dots v_{i_k})c^{m+k-1-r_o} = o$. By taking $r_o = m$ we find that for any partition ω with $|\omega| = m$, $p^*(v_{i_1} \dots v_{i_k})c^{k-1} = o$ and hence $v_{i_1} \dots v_{i_k} = o$. Since the Whitney-Stiefel classes of V^m are all o it will follow that all characteristic numbers of the bundle are o and hence $[\xi]_2 = o$ as asserted. ∎

(23.3) THEOREM: *Let* (T,M^n) *be an involution on a closed manifold with a connected and* m*-dimensional set of fixed points. If all positive dimensional tangential characteristic classes of* F^m *vanish then* $[M^n]_2 = o$.

PROOF: Since $\partial[\eta \to F^m]_2 = o$ it follows from (23.2) that $[\eta]_2 = o$. But then surely $[\eta \oplus R]_2 = o$ also and hence $[RP(\eta \oplus R)]_2 = [M^n]_2 = o$. ∎

It is not particularly difficult to see that we might have assumed only that all non-empty components of the fixed set have the same dimension and all their tangential characteristic classes vanish. The result fails if the dimensions of the components of F are allowed to vary. For example on RP(2) there is an involution whose fixed set is the disjoint union of a point with an RP(1).

24. The Smith homomorphism

We shall set up some techniques needed in the following sections. The most important is an epimorphism $\Delta : MO_n(C_2) \to MO_{n-1}(C_2)$ which we call the *Smith homomorphism*. It is defined as follows. Given a fixed point free involution on a closed manifold (T,M^n) there is for $N > n$ a smooth equivariant map $g : (T,M^n) \to (A,S^N)$ for which g is transverse regular on $S^{N-1} \subset S^N$. If we let $V^{n-1} = g^{-1}(S^{N-1})$ then V^{n-1} is a closed regular T-invariant submanifold of co-dimension 1. We claim that the correspondence $\Delta[T,M^n]_2 = [T,V^{n-1}]_2$ yields a well defined MO_*-module $\Delta : MO_*(C_2) \to MO_*(C_2)$ of degree -1.

Begin with the observation that since (A,S^N) is (N-1)-universal for principal C_2 actions there is an equivariant map $f : (T,M^n) \to (A,S^N)$ which is unique up to equivariant homotopy type. This induces $F : M^n/T \to RP(N)$. By (10.1) there is a smooth approximation $G : M^n/T \to RP(N)$ which is homotopic to F and transverse regular on $RP(N-1) \subset RP(N)$. This G is then covered by the required equivariant map $g : (T,M^n) \to (A,S^N)$ which is transverse regular on S^{N-1} since the quotient maps are local diffeomorphisms.

To show Δ is well defined for $N > n$ it is enough to show that if $[T,M^n]_2 = o$ then $[T,V^{n-1}]_2 = o$ also. Suppose $(T,M^n) = (T,\dot{B}^{n+1})$. We can extend g to an equivariant $g' : (T,B^{n+1}) \to (A,S^N)$. Using the equivariant collaring theorem there is an open invariant neighborhood $U \supset \dot{B}^{n+1}$ which is equivariantly diffeomorphic to $M^n \times [o,1]$ so we can assume that $g'(x,t) = g(x)$ on U.

There is then induced $G' : B^{n+1}/T \to RP(N)$, and $G'(x,t) = G(x)$ for $x \in M^n/T$ and $o \leq t \leq 1$. Now the restriction of G' to $M^n/T \times [o,1/2]$ is still transverse regular to $RP(N-1)$. By (10.1) there is a homotopic approximation to G', say G_1, which is transverse regular, as a map on all of B^{n+1}/T, to $RP(N-1)$ and which agrees with G on $M^n/T \times [o,1/2]$. Then G_1 is covered by the equivariant map $g_1 : (T,B^{n+1}) \to (A,S^N)$ so that if we take $W^n = g_1^{-1}(S^{N-1})$ it will follow that $(T,V^{n-1}) = (T,\dot{W}^n)$, and hence $[T,V^{n-1}]_2 = o$. The reader may show the independence of N and that Δ is an MO_*-module homomorphism. ∎

We should observe that if unoriented cobordism Whitney classes are developed, then the cobordism Whitney class $v_1 \in MO^1(M^n/T)$ of the real line bundle associated to (T,M^n) can be used to define Δ. First the cap-product $v_1 \cap [M^n/T,id]_2$ in $MO_{n-1}(M^n/T)$ is formed. Then using the classifying map $F : M^n/T \to RP(\infty)$ we let $\Delta[T,M^n]_2 = F_*(v_1 \cap [M^n/T,id]_2)$ in $MO_{n-1}(RP(\infty)) \simeq MO_{n-1}(C_2)$.

(24.1) LEMMA: *Let (T,M^n) be a smooth fixed point free involution on a closed manifold. Let $W^n \subset M^n$ be a compact regular submanifold for which $W^n \cup T(W^n) = M^n$ and $W^n \cap T(W^n) = \dot{W}^n$, then $\dot{W}^n$ is T-invariant and $\Delta[T,M^n]_2 = [T,\dot{W}^n]_2$.*

PROOF: Begin with the selection of a smooth equivariant $f : \dot{W}^n \to S^{N-1}$. Consider the normal line bundle to $\dot{W}^n$. It is obviously a product bundle. Using the invariant tubular neighborhoods of Section 20 we can find an *open* invariant tube $N \supset \dot{W}^n$ for which N is $\dot{W}^n \times (-1,1)$ and T on N is $T(x,t) = (Tx,-t)$. We may assume $\dot{W}^n \times [o,1) \subset W^n$ and $\dot{W}^n \times (-1,o] \subset T(W^n)$. Denote by $S^o \subset S^N$ the union of the north and the south poles. Then $S^N \setminus S^o$ is $S^{N-1} \times (-1,1)$ with $A(x,t) = (Ax,-t)$. Define $g : M^n \to S^N$ so that on N, $g(x,t) = (f(x),t)$ and $g\ (W^n \setminus N) =$ north pole, $g(T(W^n) \setminus N) =$ south pole. Now $g : (T,M^n) \to (A,S^N)$ is transverse regular on S^{N-1} with $g^{-1}(S^{N-1}) = \dot{W}^n$. ∎

It now follows for $n \geq 1$, that $\Delta[A,S^n]_2 = [A,S^{n-1}]_2$. Indeed there is, using the fact that $\{[A,S^n]_2\}$ is a homogeneous basis for $MO_*(C_2)$, a short exact sequence

$$o \to MO_* \to MO_*(C_2) \overset{\Delta}{\to} MO_*(C_2) \to o$$

where MO_* is embedded by $[V^n]_2 \to [A,S^o]_2[V^n]_2$.

Let us next turn to a homomorphism $I_* : MO_*(BO(k)) \to MO_*(BO(k+1))$. This is induced by the map $BO(k) \to BO(k+1)$. It sends $[\xi]_2$ to $[\xi \oplus R]_2$ for a k-plane bundle $\xi \to V^n$ over a closed manifold.

(24.2) LEMMA: *For all* $n \geq o$, $I_* : MO_n(BO(k)) \to MO_n(BO(k+1))$ *is an isomorphism whose image consists of those bordism classes of* (k+1)*-plane bundles for which every Whitney number involving* v_{k+1} *vanishes. In case* $n \leq k$ *then* $I_* : MO_n(BO(k)) \simeq MO_n(BO(k+1))$.

PROOF: This follows immediately from (8.3) and (17.4). ∎

(24.3) THEOREM: *The diagram*

$$\begin{array}{ccc} MO_n(BO(k+1)) & \overset{\partial}{\longrightarrow} & MO_{n+k}(C_2) \\ \uparrow I_* & & \downarrow \Delta \\ MO_n(BO(k)) & \overset{\partial}{\longrightarrow} & MO_{n+k-1}(C_2) \end{array}$$

is commutative.

PROOF: Let us translate the assertion into geometric language. If $\xi \to V^n$ is a k-plane bundle over a closed manifold then we want to show $\Delta[T,S(\xi \oplus R)]_2 = [T,S(\xi)]_2$ for the associated bundle involutions. Since $S(\xi \oplus R)$ is just the Whitney join of $S(\xi)$ with a trivial o-sphere bundle, the assertion will follow from (24.1). ∎

From the identification $MO_n(C_2) \simeq MO_n(RP(\infty))$ we have a Thom homomorphism $\mu : MO_n(C_2) \to H_n(RP(\infty);Z/2Z)$. This assigns to $[T,M^n]_2$ the image of the fundamental class of M^n/T under the homomorphism induced by the classifying map $F : M^n/T \to RP(\infty)$. There is also the cohomology class c in $H^1(RP(\infty);Z/2Z)$.

(24.4) EXERCISE: *For* $n \geq 1$ *the diagram*

$$\begin{array}{ccc} MO_n(C_2) & \overset{\mu}{\longrightarrow} & H_n(RP(\infty);Z/2Z) \\ \downarrow \Delta & & \downarrow \cap\, c \\ MO_{n-1}(C_2) & \overset{\mu}{\longrightarrow} & H_{n-1}(RP(\infty);Z/2Z) \end{array}$$

commutes. ∎

Now we shall discuss a right inverse $\gamma : MO_n(C_2) \to MO_{n+1}(C_2)$ to Δ. Geometrically it is defined as follows. Regard the circle S^1 as the complex numbers of modulus 1. Then the antipodal involution is $A(z) = -z$ and there is also the conjugation involution $c(z) = \bar{z}$. On $M^n \times S^1$ the involutions $T \times c$ and $id \times A$ commute so that $id \times A$ uniquely induces an involution $(t,(M^n \times S^1)/(T \times c))$. Since $T \times c$ is fixed point free the quotient $(M^n \times S^1/(T \times c)$ is a closed manifold V^{n+1}. As (A,S^1) has no fixed points and T has no coincidences with the identity it follows (t,V^{n+1}) is also fixed point free. We put $\gamma[T,M^n]_2 = [t,V^{n+1}]_2$.

Note that if $\xi \to M^n/T$ is the real line bundle associated with (T,M^n) then $V^{n+1} \to M^n$ is just the sphere bundle $S(\xi \oplus R)$ and t is actually the bundle involution. Thus γ is simply the composition

$$MO_n(C_2) \simeq MO_n(BO(1)) \xrightarrow{I_*} MO_n(BO(2)) \xrightarrow{\partial} MO_{n+1}(C_2).$$

Therefore $\Delta \circ \gamma$ = identity by (24.3). We also note, using the exercise (21.8), that $[V^{n+1}/T]_2 = [RP(\xi \oplus R)]_2 = o$. That is, $\gamma: MO_n(C_2) \to \widetilde{MO}_{n+1}(C_2)$ where the reduced module is defined to be the kernel of $\varepsilon_*[T,M^n]_2 = [M^n/T]_2$, the augmentation.

(24.5) THEOREM: *There is a unique homogeneous basis* $\{X(n)\}_o^\infty$ *of* $MO_*(C_2)$ *which satisfies*

i) $\Delta X(n+1) = X(n)$, $n \geq o$

ii) $\varepsilon_*(X(n)) = o \in MO_n$, $n > o$.

Furthermore, for any Y *in* $MO_n(C_2)$

$$Y = \textstyle\sum_o^n X(k)\varepsilon_*(\Delta^k(Y)).$$

PROOF: Suppose first that at least one such basis exists. Because it is a basis, $X(o) = [A,S^o]_2$. Now for any $Y \in MO_n(C_2)$ we can uniquely write $Y = \sum_o^n X(j)[V^{n-j}]_2$. But then $\Delta^k(Y) = \sum_k^n \varepsilon_* X(j-k)[V^{n-j}]_2$ while $\varepsilon_*(\Delta^k(Y)) = \sum_k^n \varepsilon_*(X(j-k))[V^{n-j}]_2 = [S^o/A]_2[V^{n-k}]_2 = [V^{n-k}]_2$ since $\varepsilon_*(X(j-k)) = o$ for all $j > k$.

If now $\{Y(n)\}_o^\infty$ were a second such basis, it would follow that $Y(n) = \sum_o^n X(k)\varepsilon_*(\Delta^k(Y(n)) = \sum_o^n X(k)\varepsilon_*(Y(n-k)) = X(n)$. Thus we have uniqueness.

For existence we simply put $X(n) = \gamma^n[A,S^o]_2$. As we already noted, $\varepsilon_*(X(n)) = o$ for all $n > o$ and $\Delta X(n+1) = X(n)$ for $n \geq o$. Finally, applying (24.4) inductively, we find $\mu(X(n)) \neq o \in H_n(RP(\infty);Z/2Z)$. ∎

We shall refer to this as the *natural base* for $MO_*(C_2)$. We shall point out a specific use of this base later.

25. Unrestricted bordism classes of involutions

We shall begin our study of the unoriented bordism algebra $I_*(C_2)$ of all involutions on closed manifolds. This was introduced in Section 20. First we define $M_n = \sum_0^n MO_m(BO(n-m))$. Recall that we always put $MO_n(BO(o)) = MO_n$. Using the external Whitney sum we can impose on $M_* = \sum_0^\infty M_n$ the structure of a graded commutative algebra over MO_* with identity. Specifically, given $[\xi_1 \to V_1^r]_2$ and $[\xi_2 \to V_2^s]_2$ we define the product to be

$$[\xi_1 \to V_1^r]_2[\xi_2 \to V_2^s]_2 = [\xi_1 \times \xi_2 \to V_1^r \times V_2^s]_2.$$

We note that the multiplicative identity is the o-bundle over a point. Since $MO_* = MO_*(BO(o))$, $MO_* \subset M_*$ as a subring.

Now from our discussion of characteristic classes in Section 10 it will follow, with the aid of (8.3), if $(C_2)^k \subset O(k)$ as the subgroup of diagonal matrices then the map $\rho : B((C_2)^k) \to BO(k)$ will yield an epimorphism $\rho_* : MO_*(B((C_2)^k)) \to MO_*(BO(k))$. Now $B((C_2)^k)$ is just the k-fold product of $B(C_2) = RP(\infty)$ with itself. Furthermore there is a Künneth formula from Section 19

$$MO_*(B(C_2)) \otimes_{MO_*} MO_*(BO(C_2)) \otimes \dots \otimes_{MO_*} MO_*(B(C_2)) \simeq MO_*(B((C_2)^k))$$

where we understand k-fold tensor product.

(25.1) LEMMA: *As an MO_*-algebra, M_* is a polynomial algebra with a generator in each M_n, $n > o$.*

PROOF: The easiest thing to do is simply exhibit a system of generators. For each $n > o$ let $[\xi \to RP(n-1)]_2$ in $MO_{n-1}(BO(1))$ be the real line bundle associated with (A, S^{n-1}). From our remark that $MO_*(B((C_2)^k)) \to MO_*(BO(k))$ is an epimorphism it will immediately follow that M_* is generated over MO_* by the $\left\{[\xi \to RP(n-1)]_2\right\}_1^\infty$. Let $[\xi_n]_2 = [\xi \to RP(n-1)]_2 \in M_n$. Using an argument similar to (17.3) it can be seen that there are no relations over MO_* between distinct monomials in the $[\xi_n]_2$.

A multiplicative homomorphism $j_* : I_*(C_2) \to M_*$ is given as follows. For an involution (T,M^n) put

$$j_*[T,M^n] = \sum_0^n [\eta^{n-m} \to F^m]_2 \in M_n.$$

This is well defined for if there is a (T,B^{n+1}) with $(T,M^n) = (T,\dot{B}^{n+1})$ we denote by $E^{m+1} \subset B^{n+1}$ the union of the (m+1)-dimensional components of the fixed set of (T,B^{n+1}). There are then normal bundles $\tilde{\eta}^{n-m} \to E^{m+1}$. By the Equivariant Collaring Theorem E^{m+1} is a compact manifold with $\dot{E}^{m+1} = E^{m+1} \cap \dot{B}^{n+1} = F^m$ and $\tilde{\eta}^{n-m}$ restricted to $\dot{E}^{m+1}$ agrees with $\eta^{n-m} \to F^m$. (It may happen that some components of $E \subset B^{n+1}$ are actually closed manifolds lying in the interior of B^{n+1}, but these have no effect on $\dot{E} = E \cap \dot{B}^{n+1}$). Obviously j_* is a homomorphism of MO_* algebras with degree o. There is also an MO_*-module homomorphism $\partial : M_* \to MO_*(C_2)$ with degree -1. It is the sum of the various $\partial : MO_m(BO(n-m)) \to MO_{n-1}(C_2)$ introduced in Section 23.

(25.2) THEOREM: *The sequence*

$$o \to I_*(C_2) \xrightarrow{j_*} M_* \xrightarrow{\partial} MO_*(C_2) \to o$$

is split exact.

PROOF: First we define a homomorphism $K : MO_*(C_2) \to M_*$ having degree +1 so that ∂K = identity. Any fixed point free involution (T,V^{n-1}) admits a unique decomposition

$$[T,V^{n-1}]_2 = \sum_0^{n-1} [A,S^{n-k-1}]_2 [X^k]_2.$$

Thus let

$$K([T,V^{n-1}]_2) = \sum_0^{n-1} [\xi_1]_2^{n-k} [X^k]_2 \in M_n.$$

Remember that $[\xi_1]_2^{n-k}$ is the bordism class $[(n-k)R \to pt]_2$ of the trivial $(n-k)$-plane bundle over a point. Obviously ∂K = identity.

That $\text{im}(j_*) \subset \ker(\partial)$ is exactly (22.1). To see that $\ker(\partial) \subset \text{im}(j_*)$ we proceed as follows. Suppose we have $\sum_0^n [\eta^{n-m} \to V^m]_2$. Passing to associated disk bundles we would have $(T,D(\eta)) = \sqcup_0^n (T,D(\eta^{n-m}))$ where $(T,D(\eta^{n-m}))$ is the antipodal map on each fibre. Now $(T,D(\eta))$ is an involution on a compact n-manifold with $(T,\dot{D}(\eta)) = (T,S(\eta))$ the bundle involution on the associated sphere bundle. If now $[T,S(\eta)]_2 = o \in MO_{n-1}(C_2)$ there is a fixed point free involution (T_1,B^n) on a compact manifold with $(T_1,\dot{B}^n) = (T,S(\eta))$ also. Finally then there is an involution (T_2,M^n) on the closed manifold $M^n = D(\eta) \cup B^n$, with $D(\eta) \cap B^n = S(\eta)$, for which $j_*[T_2,M^n]_2 = \sum_0^n[\eta^{n-m} \to V^m]_2$. Therefore $\mathrm{im}(j_*) = \ker(\partial)$.

The last step is to prove that j_* is a monomorphism. Thus suppose $j_*[T,M^n]_2 = \sum_0^n[\eta^{n-m} \to F^m]_2 = o$. We shall then have compact manifolds and bundles $\tilde{\eta}^{n-m} \to E^{m+1}$ with $\tilde{\eta}^{n-m}|\dot{E}^{m+1} = \eta^{n-m} \to F^m$. We introduce disk bundles with the involutions to obtain $(T,D(\tilde{\eta})) = \bigsqcup_0^n(T,D(\tilde{\eta}^{n-m}))$. With a little angle straightening each $D(\tilde{\eta}^{n-m}) \to E^{m+1}$ is a compact $(n+1)$-manifold. Furthermore

$$\dot{D}(\tilde{\eta}^{n-m}) = S(\tilde{\eta}^{n-m}) \cup D(\eta^{n-m})$$

with $S(\tilde{\eta}^{n-m}) \cap D(\eta^{n-m}) = S(\eta^{n-m}) \to F^m$. We write

$$(T,\dot{D}(\tilde{\eta})) = (T,S(\tilde{\eta})) \cup (T,D(\eta)).$$

The significant point here is that $(T,S(\tilde{\eta})) \to E$ is fixed point free. Consider now $(T',M^n \times I)$ where $T'(x,t) = (Tx,t)$. Think of $(T,D(\eta)) \subset (T,M^n \times o) \subset (T',M^n \times I)^{\cdot}$ as the closed invariant normal tube around the fixed set. Look back at $(T,\dot{D}(\tilde{\eta}))$. We form $(T,D(\tilde{\eta})) \cup (T',M^n \times I)$ by identifying the $(T,D(\eta)) \subset (T,\dot{D}(\tilde{\eta}))$ with $(T,D(\eta)) \subset M^n \times o \subset (M^n \times I)$. This new compact manifold is a bordism between the original involution and a *fixed point free* involution on $S(\tilde{\eta}) \cup (M^n \setminus \mathrm{Int}(N))$ with $S(\tilde{\eta}) \cap (M^n \setminus \mathrm{Int}(N)) = S(\eta)$.

Thus we have shown that if $j_*[T,M^n]_2 = o$ then $[T,M^n]_2$ lies in the image of the natural homomorphism $MO_*(C_2) \to I_*(C_2)$ given by forgetting the restrictions. Therefore we can assume (T,M^n) is fixed point free. Look at $(T',M^n \times I)$. If we identify (x,o) with $(T(x),o)$ we obtain an involution (T',B^{n+1}) on a compact manifold (actually the mapping cylinder of $M^n \to M^n/T$) with $F(T') = M^n/T$ and $(T',\dot{B}^{n+1}) = (T,M^n)$. This completes (25.2). ∎

In abbreviated form (25.2) says that the unrestricted bordism class of an involution is uniquely determined by the bordism class of the normal bundle to the fixed point set.

Now we take up an operator $\Gamma : I_*(C_2) \to I_*(C_2)$ which is an MO_*-module homomorphism of degree +1. This was first introduced by Jim Alexander [Ax]. Given (T,M^n) we introduce $S^1 \times M^n$ and a commuting pair of involutions

$$T_1(z,x) = (\bar{z},x); \; T_2(z,x) = (-z,Tx).$$

Since T_2 is fixed point free, the quotient $V^{n+1} = (S^1 \times M^n)/T_2$ is still a closed manifold. Because T_1 commutes with T_2 it will induce an involution (τ, V^{n+1}). We put $\Gamma([T,M^n]_2) = [\tau, V^{n+1}]_2$ and this is readily seen to define an endomorphism of degree +1 on $I_*(C_2)$.

The first step is an analysis of the fixed set of (τ, V^{n+1}) together with its normal bundle. If $\eta \to F$ is the normal bundle to the fixed set of (T,M^n) then we claim

(25.3) LEMMA: *The fixed set of* (τ, V^{n+1}) *together with its normal bundle is the disjoint union of* $\eta \oplus R \to F$ *with* $R \to M^n$.

PROOF: The fixed point set of (τ, V^{n+1}) is the quotient under T_2 of the union of the fixed point set of T_1 with the set of coincidence of T_1 and T_2. Since T_2 has no fixed points this will be a disjoint union in $S^1 \times M^n$.

The fixed point set of T_1 is $1 \times M^n \sqcup -1 \times M^n$ and under $S^1 \times M^n \to V^{n+1}$ this is carried into one copy of M^n in the fixed point set of (τ, V^{n+1}). The normal bundle of $1 \times M^n \subset S^1 \times M^n$ is the trivial line bundle of course and since a small open neighborhood of $1 \times M^n$ will map homeomorphically down in V^{n+1} we can say the normal bundle to $M^n \subset V^{n+1}$ is still $R \to M^n$.

The set of coincidence of T_1 with T_2 is $i \times F \sqcup -i \times F$ which in V^{n+1} becomes a single copy of F. The normal bundle to $i \times F \subset S^1 \times M^n$ is $\eta \oplus R \to F$ so by analogy with the preceding case $\eta \oplus R \to F$ is the normal bundle in V^{n+1}. ∎

In terms of (25.2, 24.3) we can give the formula

$$(25.4) \qquad j_*(\Gamma[T,M^n]_2) = I_*(j_*([T,M^n]_2 + [id,M^n]_2).$$

Since I_* and j_* are always monomorphisms, we find

(25.5) LEMMA: *The element* $[T,M^n]_2$ *lies in the kernel of* Γ *if and only if* $[T,M^n]_2 = [id,M^n]_2$. ∎

If we apply (25.3) and (22.2) iteratively we shall find

(25.6) LEMMA: *If* $k > 0$, *then in* MO_{n+k}

$$\varepsilon(\Gamma^k[T,M^n]_2) = [RP(\eta \oplus (k+1)R)]_2$$

$$+ \textstyle\sum_0^{k-1} (\varepsilon\Gamma^j[T,M^n]_2)[RP(k-j)]_2. \quad ∎$$

By $\varepsilon : I_*(C_2) \to MO_*$ we denote the augmentation $\varepsilon[T,M^n]_2 = [M^n]_2$. Next we want to relate Γ to the product in $I_*(C_2)$.

(25.7) LEMMA: *For any pair* x,y *in* $I_*(C_2)$

$$\Gamma(xy) = \Gamma(x)\cdot y + \varepsilon(x)\cdot\Gamma(y) = x\cdot\Gamma(y) + \varepsilon(y)\cdot\Gamma(x).$$

PROOF: We shall use the formula (25.4). The monomorphism $j_* : I_*(C_2) \to M_*$ is multiplicative. Furthermore the degree +1 monomorphism $I_* : M_* \to M_*$ is multiplication by $[\xi_1]_2 = [R \to pt]_2$.

Thus

$$j_*(\Gamma(x)\cdot y) = (j_*(x)[\xi_1]_2 + \varepsilon(x)[\xi_1]_2)(j_*(y))$$

$$j_*(\varepsilon(x)\cdot\Gamma(y)) = \varepsilon(x)(j_*(y)[\xi_1]_2 + \varepsilon(y)[\xi_1]_2)$$

while

$$j_*(\Gamma(xy)) = j_*(x)j_*(y)[\xi_1]_2 + \varepsilon(x)\varepsilon(y)[\xi_1]_2.$$

Multiply out the first two lines, add them together and $\Gamma(xy) = \Gamma(x)\cdot y + \varepsilon(x)\cdot\Gamma(y)$. By a similar consideration $\Gamma(xy) = x\Gamma(y) + \Gamma(x)\cdot\varepsilon(y)$ also. ∎

As a consequence of (25.7) we can point out that since $\varepsilon(\varepsilon(y)) = \varepsilon(y)$, $\varepsilon(\varepsilon(x)) = \varepsilon(x)$.

(25.8) LEMMA: *The composite* $\varepsilon\circ\Gamma : I_*(C_2) \to MO_*$ *satisfies*

$$(\varepsilon\circ\Gamma)(xy) = ((\varepsilon\circ\Gamma)(x))\varepsilon(y) + \varepsilon(x)((\varepsilon\circ\Gamma)(y)).\quad ∎$$

26. Stabilizing

The algebra $I_*(C_2)$ is quite complicated. Following Royster's idea we shall introduce a quotient algebra which has a stabilizing effect and which is easy to identify $[Ro_2]$.

(26.1) LEMMA: *If* $S \subset I_*(C_2)$ *is defined by*

$$S = \{x + \Gamma(x) \mid \varepsilon(x) = o \in MO_*\}$$

then S *is an ideal.*

PROOF: Consider any element $y \in I_*(C_2)$ Surely $\varepsilon(xy) = o$. According to (25.7), since $\varepsilon(x) = o$, $\Gamma(xy) = \Gamma(x)y$ and thus $xy + \Gamma(xy) = (x + \Gamma(x))y$, which is as required. Obviously S is a subgroup. ∎

The quotient algebra $I_*(C_2)/S$ will be denoted by $\Lambda(C_2)$. This is no longer a graded algebra. In fact the relation of S to homogeneous elements is

(26.2) LEMMA: *For any* $k \geq o$, $S \cap I_k(C_2) = o$.

PROOF: Suppose $x = (x_m, \ldots, x_n)$ with $x_m \neq o$, $x_n \neq o$ and $\varepsilon(x) = o$, $x + \Gamma(x) \in I_k(C_2)$. A moment's reflection shows that $x_m = x_n = x_k$ with $\Gamma(x_k) = o$. But then by (25.5) $x_k = \varepsilon(x_k) = o$. ∎ We state explicitly the corollary

(26.3) LEMMA: *For any* $k \geq o$

$$I_k(C_2) \to \Lambda(C_2)$$

is a monomorphism.

It is clear that we should factor M_* by some kind of corresponding ideal If $\varepsilon(x) = o$ then $j_*(\Gamma(x)) = j_*(x)[\xi_1]_2$ and $j_*(x) + j_*(\Gamma(x)) = j_*(x)([\xi_1]_2 + 1)$. Therefore we need the quotient of M_* by the principal ideal $([\xi_1]_2 + 1)$. In other words we need to introduce bordism classes of stable vector bundles over closed manifolds; that is, we pass to $MO_*(BO)$. In so doing we identify $\xi \to V^n$ with $\xi \oplus R \to V^n$ for any k-plane bundle over a closed manifold. Thus $M_*/([\xi_1]_2 + 1)$ is $MO_*(BO)$. Again we caution that $M_*/([\xi_1]_2 + 1)$ does not receive a grading from M_* because degree in M_* is given by adding the dimension of the fibre to the dimension of the base. However, it is certainly true that $j_* : I_*(C_2) \to M_*$ will induce $j_* : \Lambda(C_2) \to MO_*(BO)$.

(26.4) THEOREM: *The induced* $j_* : \Lambda(C_2) \to MO_*(BO)$ *is an isomorphism.*

PROOF: If the proof of (25.2) is examined we see that to show the induced $\Lambda(C_2) \to MO_*(BO)$ is an epimorphism it is sufficient to show that modulo the ideal $(1 + [\xi_1]_2)$ every element of the form $\sum_o^{n-1}[\xi_1]_2^{n-k}[X^k]_2$ lies in the image of $j_* : I_*(C_2) \to M_*$. However, for $n - k > o$

$$1 + [\xi_1]_2^{n-k} = (1 + [\xi_1]_2)(1 + [\xi_1]_2 + \ldots + [\xi_1]_2^{n-k-1})$$

from which it follows

$$\sum_o^{n-1}[\xi_1]_2^{n-k}[X^k]_2 = \sum_o^{n-1}[X^k]_2$$

modulo $(1 + [\xi_1]_2)$. Now $[X^k]_2 \in MO_k(BO(o)) \subset M_k$ is the image of $[id, X^k] \in I_k(C_2)$.

Next we must show that $\Lambda(C_2) \to MO_*(BO)$ has a trivial kernel. Thus we have an $x \in I_*(C_2)$ for which we can write

$$j_*(x) = \alpha + I_*(\alpha) = \alpha(1 + [\xi_1]_2)$$

for some $\alpha \in M_*$. Let us write out $\alpha = (\alpha_m, \ldots, \alpha_n)$ with $\alpha_k \in M_k$ and $\alpha_n \neq o$, $\alpha_m \neq o$. Then $j_*(x_k) = \alpha_k + I_*(\alpha_{k-1})$. Of course $x_k = o$ if $k < m$ or $k > n + 1$. Notice that in $MO_*(C_2)$ we must have $\partial(\alpha + I_*(\alpha)) = o$ Then from dimensional considerations we must have $\partial I_*(\alpha_n) = o$ and $\partial(\alpha_n) = \partial I_*(\alpha_{n-1})$. But using (24.3), $\partial(\alpha_n) = o = \partial I_*(\alpha_{n-1})$. Proceeding inductively we establish that $\partial(\alpha_k) = o$ and $\partial(I_*(\alpha_k)) = o$ for all k. Surely then there is a $y_k \in I_k(C_2)$ with $j_*(y_k) = \alpha_k$. On the other hand, $j_*\Gamma(y_k) = I_*(\alpha_k) + \varepsilon(y_k)[\xi_1]_2$. Applying ∂ we find $\partial I_*(\alpha_k) + [A, S^o]_2 \in (y_k) = o$ in MO_{k-1} and we just saw $\partial I_*(\alpha_k) = o$. Hence we conclude that $\varepsilon(y_k) = o$. So $y_k + \Gamma(y_k) \in S$ and with $y = (y_m, \ldots, y_n)$ we find $x = y + \Gamma(y)$ as required. ∎

Evidently $MO_*(BO)$ is still a polynomial ring over MO_*. One possible choice of generators is given by the twisted real line bundles $\xi \to RP(n)$, $n \geq 1$. Frankly this is not particularly useful for purposes. The right viewpoint for us is to regard $\Lambda(C_2)$ as a polynomial algebra over $Z/2Z$ and to filter it.

(26.5) DEFINITION: *If* $x \in \Lambda(C_2)$ *then* $\mathrm{Fil}(x) \leq n$ *if and only if* $j_*(x) \in \sum_o^n MO_k(BO)$.

This is an increasing filtration. Obviously we have correspondingly filtered $MO_*(BO)$. If (T, M^m) is an involution on a closed manifold for which every non-empty component of the fixed point set has dimension $\leq n$ then the filtration of the corresponding element in $\Lambda(C_2)$ is $\leq n$.

Our filtration satisfies $F_n \cdot F_k \subset F_{n+k}$. In view of this we may say that $x \in F_n$ is $Z/2Z$-*decomposable* if and only if x can be expressed as a sum of products of elements of lower filtration. Otherwise x is $Z/2Z$-indecomposable.

For the moment forget $\Lambda(C_2)$ and consider $MO_*(BO)$ with $F_n = \sum_o^n MO_k(BO)$. The question is how to recognize a system of generators of $MO_*(BO)$ as a polynomial algebra over $Z/2Z$. We shall need the augmentation $\varepsilon_* : MO_*(BO) \to MO_*$.

(26.6) LEMMA: *Let* $\alpha_n \in F_n$ *denote a sequence of elements defined for all* $n \geq 1$ *such that*

(i) *the* α_n *generate* $MO_*(BO)$ *as a polynomial algebra over* MO_*

(ii) *for each* n *the augmentation* $\varepsilon_*(\alpha_n)$ *in* $\sum_0^n MO_k$ *is* Z/2Z-*decomposable. Let* β_n *be a sequence of elements,* $\beta_n \in F_n$, *defined for each* n *not of the form* $2^j - 1$ *and such that the* $\varepsilon_*(\beta_n)$ *generate* MO_* *as a* Z/2Z *polynomial algebra. The conclusion is that taken all together the* α_n, β_n *generate* $MO_*(BO)$ *as a polynomial algebra over* Z/2Z.

PROOF: It is immediately clear that $\alpha_n, \varepsilon_*(\beta_n)$ do generate $MO_*(BO)$ as a Z/2Z-polynomial algebra. The issue is showing that β_n can replace $\varepsilon_*(\beta_n)$. We can choose p,q in Z/2Z so that

$$\beta_n = p\varepsilon_*(\beta_n) + q\alpha_n + Z/2Z\text{-decomposables.}$$

Applying ε_* we have

$$\varepsilon_*(\beta_n) = p\varepsilon_*(\beta_n) + q\varepsilon_*(\alpha) + \text{decomposables.}$$

Since $\varepsilon_*(\alpha_n)$ is decomposable it will follow that $p = 1 \in Z/2Z$ and therefore

$$\varepsilon_*(\beta_n) = \beta_n + q\alpha_n + Z/2Z\text{-decomposables.}$$

Now α_n, β_n generate Z/2Z as required. ∎

While we do know how to recognize the $\varepsilon_*(\beta_n)$ as polynomial ring generators of MO_* we shall have to discuss the α_n a bit. Suppose $\xi \to V^n$ is a k-plane bundle over a closed manifold. We factor the total Whitney class of the bundle into $(1 + t_1)\ldots(1 + t_k)$ and we have

$$S_n(\xi) = \langle t_1^n + \ldots + t_k^n, \sigma_n\rangle \in Z/2Z.$$

Obviously this is stable in the sense that $s_n(\xi \oplus R) = s_n(\xi)$. It is a bordism invariant and so yields a homomorphism $s_n : MO_n(BO) \to Z/2Z$. We leave it to the reader to show that if $n > o$, $m > o$, then for a product $\xi \times \xi' \to V^n \times M^m$ the value $S_{n+m}(\xi \times \xi')$ is o in Z/2Z. Thus the invariant detects indecomposability in $MO_*(BO)$. Clearly for $\alpha_n \in F_n = \sum_0^n MO_j(BO)$ we can speak of $s_n(\alpha_n)$ referring to the coordinate of α_n in the summand $MO_n(BO)$. By writing $s_n(\varepsilon_*(\alpha_n))$ we mean to evaluate the invariant on the stable tangent bundle to the base manifold of the coordinate of α_n in $MO_n(BO)$. We close the section with

(26.7) LEMMA: *Let* α_n, β_n *be two sequences,* $n \geq 1$*, of elements with* α_n, β_n *in* F_n *and* β_n *omitted if* $n = 2^j - 1$*. If* $s_n(\alpha_n) = 1$, $s_n(\varepsilon_*(\alpha_n)) = o$ *and* $s_n(\varepsilon_*(\beta_n)) = 1$ *then* α_n, β_n *generate* $MO_*(BO)$ *as a polynomial algebra over* $Z/2Z$. ∎

27. The Boardman theorems

We are now prepared to reproduce the results about involutions due to Boardman [Ba]. Here now is the key lemma.

(27.1) LEMMA: *There is a sequence of involutions on closed manifolds* $(T,Y(n))$, $n \neq 2^j - 1$, *such that*

(i) $[Y(n)]_2$ *is indecomposable in* MO_n.

(ii) *if* $n = 2k$ *then every component of* F *has dimension* $\leq k$, $s_k(\eta \to F) = 1$ *and* $[F^k]_2$ *is decomposable*

(iii) *if* $n = 2k + 1$ *every component of* F *has dimension* $\leq k$ *and* $s_k([F^k]_2) = 1$.

PROOF: This will necessarily be of a constructive nature and is divided into several parts. First there is $n = 4m + 2$ and we can use the involution on $RP(4m + 2)$ given by

$$T[x_1,\ldots,x_{4m+3}] = [x_1,\ldots,x_{2m+2},-x_{2m+3},\ldots,-x_{4m+3}].$$

The fixed point set F is the disjoint union $RP(2m + 1) \sqcup RP(2m)$. Furthermore the normal bundle to $RP(2m + 1)$ is the Whitney sum $(2m + 1)\xi \to RP(2m + 1)$. Thus the total Whitney class is $(1 + c)^{2m+1}$ and hence $s_{2m+1}[(2m + 1)\xi \to RP(2m + 1)]_2 = (2m + 1) \langle c^{2m+1}\sigma\rangle = 1$. Of course $[RP(2m + 1)]_2 = o$.

Next consider $n = 4m$. Begin with the involution $(d,RP(2m) \times RP(2m))$ which switches coordinates. The fixed point set is $RP(2m) = \Delta \subset RP(2m) \times RP(2m)$. The normal bundle to this fixed set is equivalent to the tangent bundle of $RP(2m)$ so that $s_{2m}([\eta \to RP(2m)]_2) = 1$. Now we introduce $(\tau,RP(4m))$ by $\tau[x_1,\ldots,x_{4m+1}] = [x_1,\ldots,x_{2m+1},-x_{2m+2},\ldots,-x_{4m+1}]$ so that the fixed set is $RP(2m) \sqcup RP(2m - 1)$. But this time the normal bundle to $RP(2m)$ is $2m\xi \to RP(2m)$ and hence $s_{2m}[2m\xi \to RP(2m)]_2 = o$. Our choice then will be the disjoint union

$$(T,Y(4m)) = (d,RP(2m) \times RP(2m)) \sqcup (\tau,RP(4m)).$$

We turn now to n odd and not of the form $2^j - 1$. We write $n = 2^{k+1}m + 2^k - 1$ with $k > o$, $m > o$. We shall use an involution on the manifold $RP(2^k, 2^{k+1}m - 1)$ which was introduced in Section 21. The manifold has an indecomposable bordism class. We can regard $RP(2^k, 2^{k+1}m - 1)$ as the quotient of the fixed point free involution on $S^{2^k} \times RP(2^{k+1}m - 1)$ given by

$$T_2(x_1, \ldots, x_{2^k+1}, [y_1, \ldots, y_{2^{k+1}m}]) =$$
$$(-x_1, \ldots, -x_{2^k+1}, [-y_1, y_2, \ldots, y_{2^{k+1}m}]).$$

This commutes with

$$T_1(x_1, \ldots, x_{2^k+1}, [y_1, \ldots, y_{2^{k+1}m}]) =$$
$$(x_1, \ldots, x_{2^{k+1}+1}, -x_{2^{k-1}+2}, \ldots, -x_{2^k+1}, [y_1, \ldots, y_{2^km}, -y_{2^km+1}, \ldots, -y_{2^{k+1}m}])$$

so that T_1 will induce $(T, RP(2^k, 2^{k+1}m - 1))$.

To see the fixed point set of T we must exhibit the fixed point set of T_1 and the set of coincidences of T_1 and T_2. In $F(T_1)$ we find exactly a disjoint union of $S^{2^{k-1}} \times RP(2^km - 1)$ and another copy of $S^{2^{k-1}} \times RP(2^km - 1)$. But in passing to the quotient under T_2 these become respectively $RP(2^{k-1}, 2^km - 1)$ and $RP(2^{k-1}) \times RP(2^km - 1)$.

The coincidences of T_1 and T_2 are given by

$$\{(o, \ldots, o, x_{2^{k-1}+2}, \ldots, x_{2^k+1}, [o, y_2, \ldots, y_{2^km}, o, \ldots, o])\} =$$
$$S^{2^{k-1}-1} \times RP(2^km - 2)$$

together with the disjoint

$$\{(o, \ldots o, x_{2^{k-1}+2}, \ldots, x_{2^k+1}, [y_1, o, \ldots, o, y_{2^km+1}, \ldots, y_{2^{k+1}m}])\} =$$
$$S^{2^{k-1}-1} \times RP(2^km).$$

And in $RP(2^k, 2^{k+1}m - 1)$ these respectively become $RP(2^{k-1} - 1) \times RP(2^km - 2)$ and $RP(2^{k-1} - 1, 2^km)$.

As long as $k > 1$ it follows that $[RP(2^{k-1}, 2^km - 1)]_2$ remains indecomposable while $[RP(2^{k-1} - 1, 2^km]_2$ is seen to be decomposable by (21.6). In case $k = 1$, however, $[RP(1, 2m - 1)]_2 = o$ by (21.8) while $[RP(o, 2m)]_2 = [RP(2m)]_2$ is indecomposable. ∎

Let $y(n) \in \Lambda(C_2)$ denote the stable bordism class determined by $(T,Y(n))$. Then $\mathrm{Fil}(y(2m))$ and $\mathrm{Fil}(y(2m+1)) = m$ if $2m+1 \neq 2^j - 1$.

(27.2) LEMMA: *The elements* $y(n)$ *generate* $\Lambda(C_2)$ *as a polynomial ring over* $Z/2Z$. *If* $x_1, x_2 \in \Lambda(C_2)$ *then* $\mathrm{Fil}(x_1 \cdot x_2) = \mathrm{Fil}(x_1) + \mathrm{Fil}(x_2)$. *If* x_1 *and* x_2 *are expressed as polynomials in the* $y(n)$ *and if they have no common monomials, then* $\mathrm{Fil}(x_1 + x_2) = \max(\mathrm{Fil}\,x_1, \mathrm{Fil}\,x_2)$. ∎

Our problem now is the lack of a natural augmentation homomorphism of $\Lambda(C_2)$ into MO_*. It is not possible to get a commutative diagram

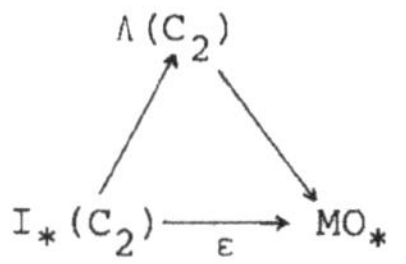

We must find a way around this difficulty. To do so we introduce $MO_*[[t]]$, the ring of homogeneous formal power series in one variable over the graded ring MO_*. An element of $MO_*[[t]]$ is a power series of the form $\sum_0^\infty [X^r]_2 t^r$ where $[X^r]_2 \in MO_r$. A ring homomorphism $E : I_*(C_2) \to MO_*[[t]]$ is defined by

$$E([T,M^n]_2) = \sum_0^\infty \varepsilon\, \Gamma^k([T,M^n]_2)\, t^{n+k}.$$

That this correspondence is multiplicative follows from (25.8) immediately. Furthermore if $[M^n]_2 = 0$ we see that $E([T,M^n]_2) = E(\Gamma([T,M^n]_2))$ so that in fact there is induced $E : \Lambda(C_2) \to MO_*[[t]]$. Note in particular $E(y(n)) = [Y(n)]_2 t^n$ + higher terms in t and that $[Y(n)]_2$ is indecomposable.

There is a natural descending sequence of ideals $J(k) \supset J(k+1) \supset \ldots$ in $MO_*[[t]]$ where

$$J(k) = \{\textstyle\sum_0^\infty [X^r]_2 t^r \mid [X^0]_2 = \ldots = [X^{k-1}]_2 = 0\}.$$

We may therefore introduce a second filtration of $\Lambda(C_2)$ by $\mathrm{fil}(x) \geq k$ if and only if $E(x) \in J(k)$. This is a decreasing filtration and for each $y(n)$, $\mathrm{fil}(y(n)) = n$. We leave to the reader

(27.3) LEMMA: *If* x_1, x_2 *are in* $\Lambda(C_2)$ *then* $\mathrm{fil}\,x_1 \cdot x_2 = \mathrm{fil}\,x_1 + \mathrm{fil}\,x_2$. *If* x_1 *and* x_2 *are expressed as polynomials in the* $y(n)$ *and if they have no common monomials then* $\mathrm{fil}(x_1 + x_2) = \min(\mathrm{fil}(x_1), \mathrm{fil}(x_2))$. ∎

Evidently we have shown $E : \Lambda(C_2) \to MO_*[[t]]$ is a monomorphism and as a consequence the composition $I_n(C_2) \to \Lambda(C_2) \to MO_*[[t]]$ is also a monomorphism for each $n \geq 0$. The least positive rational number which satisfies $\mathrm{fil}\, y(n) \leq N\, \mathrm{Fil}\, y(n)$ for all n is 5/2. If x is a monomial in the y(n) then using the multiplicative formulas in (27.2) and (27.3) we see $\mathrm{fil}\, x \leq 5/2\ \mathrm{Fil}\, x$. Finally, an arbitrary $x \in \Lambda(C_2)$ can be uniquely expressed as a sum of distinct monomials. But fil(x) is then the filtration of that monomial which minimizes fil, while Fil(x) is the filtration of that monomial which maximizes Fil. Therefore we have shown

(27.4) BOARDMAN: *For every non-trivial element* $x \in \Lambda(C_2)$

$$\mathrm{fil}(x) \leq 5/2\ \mathrm{Fil}(x). \quad \blacksquare$$

All we need do now is draw consequences of this theorem. Suppose (T,M^n) is an involution on a closed manifold. By dim F we mean the dimension of the highest dimensional non-empty component of the fixed point set. Obviously $\dim F \geq \mathrm{Fil}[T,M^n]_2$. Furthermore, if $[T,M^n]_2 \neq o \in I_n(C_2)$ it determines a non-zero element in $\Lambda(C_2)$ by (26.3) and $n \leq \mathrm{fil}[T,M^n]_2$.

(27.5) COROLLARY: *If* (T,M^n) *is an involution on a closed manifold for which* $[T,M^n]_2 \neq o$ *then* $n \leq 5/2 \dim F$. ■

We could also mention the following in passing

(27.6) COROLLARY: *If* (T,M^n) *is an involution on a closed manifold for which* $[T,M^n]_2 \neq o$ *then there is an integer* k, $o \leq k \leq 3n/2$, *for which* $\varepsilon\Gamma^k([T,M^n]_2) \neq o \in MO_{n+k}$. ■

(27.7) COROLLARY: *If* (T,M^n) *is an involution on a closed manifold for which* $[M^n]_2 \in MO_n$ *is indecomposable then*

$$\dim F \geq [n/2].$$

PROOF: Let $x \in \Lambda(C_2)$ correspond to $[T,M^n]_2$. Using E we see that y(n) must appear as a monomial of least fil in the expression for x as a polynomial. But $\mathrm{Fil}\, y(n) = [n/2]$ so (27.7) follows from (27.2). ■

The Boardman results confirm explicitly that a non-bounding involution cannot have a fixed set which is too low dimensional. In the first edition of this tract some evidence was presented indicating this was the case and Boardman's work completely answered all of our questions about this point. We see here an example of what we feel is the most important development in the bordism study of finite groups since the appearance of the first edition; namely, the evolution of techniques for dealing effectively with the bordism groups of unrestricted group actions on closed manifolds in both the oriented and unoriented cases. We remind the reader to consult [KSt].

28. The mod 2 Euler characteristic

We shall use the formulation in Section 27 to prove

(28.1) THEOREM: *If* (T,M^{2n}) *is an involution on a closed manifold of odd Euler characteristic then* $n \leq \dim F$.

The Euler characteristic reduced mod 2 of a closed manifold is a bordism invariant because it is given by the Stiefel-Whitney number $\langle w_n, \sigma_n \rangle \in Z/2Z$ For an odd dimensional manifold this is always zero. For a finite complex, X, we understand explicitly that

$$\chi(X) = \sum (-1)^i \dim H_i(X; Z/2Z).$$

Furthermore, if $\xi \to X$ is a k-plane bundle, $k > o$, over X then $\chi(RP(\xi)) = \chi(X)\chi(RP(k-1))$. This is clear for, as we commented earlier, the mod 2 homology of $RP(\xi)$ is additively isomorphic to that of the cartesian product $X \times RP(k-1)$. Let us now prove a lemma due originally to P.A. Smith.

(28.2) LEMMA: *If* (T,M^n) *is an involution on a closed manifold then*

$$\chi(M^n) \equiv \chi(F) \pmod 2.$$

PROOF: We shall assume $F^n = \emptyset$ for this causes no loss of generality. There are then two cases.

If n is odd we must show $\chi(F) \equiv o \pmod 2$. We appeal to (22.1) to see $RP(\eta)$ has 0 Euler characteristic mod 2 since $[RP(\eta)]_2 = o$. Therefore $\chi(RP(\eta)) = \sum_0^{n-1} \chi(F^m)\chi(RP(n-m-1)) \equiv \sum_{m \text{ even}} \chi(F^m) \equiv o \pmod 2$ as required.

On the other hand if n is even we write $[M^n]_2 = [RP(\eta \oplus R)]_2$ by (22.2). Then $\chi(RP(\eta \oplus R)) = \sum_0^{n-1} \chi(F^m)\chi(RP(n-m)) \equiv \sum_{m \text{ even}} \chi(F^m)\chi \pmod 2$. ∎

With a change of notation let $\chi : I_*(C_2) \to Z/2Z$ denote the ring homomorphism defined by $[T,M^n]_2 \to \langle w_n, \sigma_n \rangle$. This will vanish on the ideal S for when $[T,M^n]_2 + \Gamma[T,M^n]_2$ belongs to S then by definition $[M^n]_2 = o$ which implies $\chi(F) \equiv o \pmod 2$. But the fixed set of $\Gamma[T,M^n]_2$ is $F \sqcup M^n$ so that $\chi(\varepsilon\Gamma[T,M^n]_2) \equiv o \pmod 2$ also. In this fashion we induce a ring homomorphism $\chi : \Lambda(C_2) \to Z/2Z$.

If the reader will refer back to our description of the $[T,Y(n)]_2$ in (27.1) he will find $\chi(y(n)) = 1 \in Z/2Z$ if and only if $n \equiv 2 \pmod 4$. Now given $[T,M^{2n}]_2$ with $\langle w_{2n}, \sigma_{2n}\rangle = 1 \in Z/2Z$ we let x be the corresponding element of $\Lambda(C_2)$. In expressing x as a polynomial there will have to be at least one monomial, call it y, in the $y(4m+2)$. Now $\mathrm{fil}(x) = 2n \leq \mathrm{fil}(y)$, while $\mathrm{Fil}(x) \geq \mathrm{Fil}(y) = 1/2\ \mathrm{fil}(y) \geq n$. Since $\dim F \geq \mathrm{Fil}(x)$ this completes the proof of (28.1). ∎

This theorem can also be proved in a non-differentiable setting $[Br_1]$. We can also state

(28.3) LEMMA: *Let* (T,M^{2n}) *be an involution on a closed manifold of odd Euler characteristic then there is an* m, $n \leq m \leq 2n$, *for which* $[\eta^{2n-m} \to F^m]_2$ *is not in the image of*

$$I_* : MO_m(BO(2n-m-1)) \to MO_m(BO(2m-n)).$$

PROOF: Suppose to the contrary that each $[\eta^{2n-m}]_2$ does lie in the image of I_*. Then certainly $[F^{2n}]_2 = o$ so that we may as well assume $F^{2n} = \emptyset$. Thus we have bundles $\xi^{2n-m-1} \to \tilde{F}^m$, $o \leq m < 2n$ for which

$$[\xi^{2n-m-1} \oplus R \to \tilde{F}^m]_2 = [\eta^{2n-m} \to F^m]_2.$$

But $\partial[\eta \to F]_2 = o \in MO_{2n-1}(C_2)$ and by (24.3) $\partial[\xi \to \tilde{F}]_2 = o \in MO_{2n-1}(C_2)$ also. Therefore there is an involution (T,V^{2n-1}) with fixed set F. This contradicts (28.2) however, because $\chi(\tilde{F}) \equiv \chi(F) \equiv 1 \pmod 2$ while $\chi(V^{2n-1}) \equiv o \pmod 2$.

Incidentally, if $m < n$ then $I_* : MO_m(BO(2n-m-1) \simeq MO_m(BO(2n-m))$ according to (24.2) so we know the value of m in question satisfies $n \leq m \leq 2n$. Also if $F^{2n} = \emptyset$ we know that $v_{2n-m} \neq o \in H^{2n-m}(F^m;Z/2Z)$ for at least one value in the range $n \leq m < 2n$. ∎

Let us use this to prove a special result.

(28.4) THEOREM: *Let* (T,M^n), $n > o$, *be an involution on a closed manifold whose fixed point set is the disjoint union of a point and a* k*-sphere. Then* k = 1,2,4 *or* 8; n = 2k, *and* M^n *is mod* 2 *bordant to the appropriate projective plane.*

PROOF: Since no involution of a positive dimensional manifold can have exactly an odd number of fixed points it follows that $k > o$ and that the isolated fixed point together with the S^k lie in the same component of M^n. The involution is fixed point free on the complement of that component so we may as well assume M^n is connected and $F^n = \emptyset$.

Now the fixed set has odd Euler characteristic and therefore so does M^n. Hence n is even. To apply (28.3) the only candidate is the k-sphere. Hence $n/2 \leq k < n$ and $v_{n-k} \neq o \in H^{n-k}(S^k;Z/2Z)$. Obviously $n-k = k$ or $2k = n$.

Thus we have a normal k-plane bundle $\eta \to S^k$ with $v_k \neq o \in H^k(S^k;Z/2Z)$. But Milnor shows that k = 1, 2, 4 or 8 in that case [M_4].

Finally on each of the projective planes RP(2), CP(2), QP(2) and the Caley plane there is an involution whose fixed point set is the disjoint union of a point and the appropriate S^k. Furthermore the normal bundle to the sphere in each of these cases has $v_k \neq o \in H^k(S^k;Z/2Z) \simeq Z/2Z$. Using this the reader can show that in $I_{2k}(C_2)$ the involution in (28.4) is bordant to the standard involution on the suitable projective plane. ■

29. The RP(2r) as the fixed point set

The purpose of this section is a proof of

(29.1) STONG: *If* (T,M^n) *is a non-trivial involution on a closed manifold whose fixed point set is diffeomorphic to* RP(2r) *then* $n = 4r$ *and* $[M^n]_2 = [RP(2r)]_2^2$.

This result by R.E. Stong [St_1] completely answered a question posed in the first edition. Let us begin with a lemma which suggested the problem.

(29.2) LEMMA: *If* (T,M^n) *is a non-trivial involution on a closed manifold with fixed set* RP(2r) *then* $n = 4r$ *and the total Whitney class of* $\eta \to RP(2r)$, *the normal bundle, is* $(1+d)^m$ *where* $m \geq 2r$ *is odd,* $\binom{m}{2r} = 1$ mod 2 *and* d *is the generator of* $H^1(RP(2r);Z/2Z)$.

PROOF: From the structure of the Grothendieck ring KO(RP(2r)) we know that the total class of $\eta \to RP(2r)$ is $(1+d)^m$ for some integer $m \geq o$. Furthermore, $\chi(RP(2r)) = 1$ so that $\chi(M^n) \equiv 1 \pmod 2$, hence n is even and by (28.1) $n \leq 4r$. Let $k = n - 2r$ be the fibre dimension of the normal bundle so that k is even and $o < k \leq 2r$. Using (28.3) we see that $v_k = \binom{m}{k}d^k \neq o$ and hence $m \geq k$.

We claim that m is odd and thus $m > k$. Suppose on the contrary that m were even. There is the bundle involution $(T,S(\eta))$ associated with η and $[T,S(\eta)]_2 = o \in MO_{n-1}(C_2)$. The Stiefel-Whitney class W_1 of $S(\eta)/T = RP(\eta)$ of the tangent bundle is found from (21.4) to be

$$W_1 = p^*(w_1) + p^*(v_1) + \binom{k}{1}c = p^*(d) + mp^*(d) + kc = p^*(d)$$

since m is assumed even. However $W_1^{2r}c^{k-1} = p^*(d^{2r})c^{k-1} \neq o$, since $d^{2r} \neq o$, and this would imply that the involution number $\langle W_1^{2r}c^{k-1},\sigma\rangle \neq o$ for $[T,S(\eta)]_2 = o$. From this contradiction we find that m is odd and larger than k.

It must also follow that $m > 2r$ for if $2r \geq m > k$ we would have $v_m \neq o$, contradicting the fact that η is a k-plane bundle.

Finally we wish to show $k = 2r$ and $\binom{m}{2r} \equiv 1 \pmod 2$. Since $v_k \neq o$ and $v_k = \binom{m}{k}d^k$ we have $\binom{m}{k} \equiv 1 \pmod 2$. Now m is odd and k is even so it must follow that $\binom{m}{k+1} \equiv 1 \pmod 2$ also. Thus if $k < 2r$ we would have the contradiction $v_{k+1} = \binom{m}{k+1}d^{k+1} \neq o$ for a k-plane bundle. ∎

We shall continue now with Stong's argument. From (21.5) the total Whitney-Stiefel class of the tangent bundle to $RP(\eta)$ is

$$(1 + p^*(d))^{2r+1}(1 + c + p^*(d))^m.$$

Now since m is odd we can write either $m = 2r + 1$, in which case we are done, or $m = 2r + 1 + 2^s(2k + 1)$ with $s \geq 1$ and $k \geq o$. We must then show $2r < 2^s$. If we put $\alpha = p^*(d)$ then the Whitney-Stiefel class can be re-expressed as

$$(1+c+\alpha(\alpha+c))(1+c^2+\alpha^2(\alpha^2+c^2))(1+\alpha^{2^s}+c^{2^s})(1+\alpha^{2^{s+1}}+c^{2^{s+1}})^k.$$

First we show that $s > 1$. Suppose $s = 1$, then we immediately read off

$$W_2 = c\alpha + (r+1)c^2$$

or

$$c\alpha = W_2 + (r+1)c^2.$$

Consider then

$$c^{2r}\alpha^{2r-1} = c(c\alpha)^{2r-1} = c(W_2 + (r+1)c^2)^{2r-1}.$$

Clearly this is an involution number of $[T,S(\eta)]_2$ and so it should vanish. On the other hand, from (21.3),

$$c^{2r} = c^{2r-1}\alpha + \text{terms involving higher powers of } \alpha.$$

Therefore $c^{2r}\alpha^{2r-1} = c^{2r-1}\alpha^{2r} \neq o$, a contradiction.

We have found $s > 1$ and hence

$$W_2 = \alpha^2 + c\alpha + rc^2$$

or

$$c\alpha + \alpha^2 = W_2 + rc^2.$$

Using this we again rewrite the total Whitney-Stiefel class of RP(η) into the form

$$(1+c+W_2+rc^2)(1+\alpha^{2^s}+c^{2^s})(1+c^2+W_2^2+rc^4)^r(1+\alpha^{2^{s+1}}+c^{2^{s+1}})^k.$$

As an immediate consequence

$$W_{2^s} = \alpha^{2^s} + P(W_2,c)$$

where $P(W_2,c)$ is some polynomial, of appropriate degree in W_2 and c.

If we assume $2r \geq 2^s$ we can write $2r = p2^s + q$ $(p \geq 1,\ o \leq q < 2^s)$. Then we assert

$$(\alpha(c+\alpha))^q\, \alpha^{p2^s} c^{2r-1-q} = \alpha^q c^q \alpha^{p2^s} c^{2r-1-q} = \alpha^{2r} c^{2r-1}$$

because all the other terms in $(\alpha(c+\alpha))^q \alpha^{p2^s}$ contain a power of α larger than 2r and hence vanish. Therefore

$$c^{2r-1}\alpha^{2r} = (W_2 + rc^2)^q (W_{2^s} + P(W_2,c))^q c^{2r-1-q}$$

which is obviously an involution number and so should vanish in contradiction to the fact that $\alpha^{2r} \neq o$.

We have now shown that $2^s > 2r$ and, since $m = 2r+1+2^s(2k+1)$,

$$(1+d)^m = (1+d)^{2r+1}(1+d^{2^s})^{2k+1} = (1+d)^{2r+1}$$

since $d^{2^s} = o$. Thus the normal bundle $\eta \to RP(2r)$ has the same characteristic class as does the tangent bundle to RP(2r). Hence $[\eta \to RP(2r)]_2 = [\tau \to RP(2r)]_2$ in $MO_{2r}(BO(2r))$. Finally, from (25.1) it will follow that $[T,M^{4r}]_2 = [d,RP(2r) \times RP(2r)]_2$ in $I_{4r}(C_2)$. ∎

30. Aspherical generators for MO_*

Our effort now will be directed at proving

(30.1) THEOREM: *There is a sequence of closed aspherical manifolds whose unoriented bordism classes generate* MO_* *as a* Z/2Z-*polynomial algebra.*

We have borrowed the classical term *aspherical* from knot theory to describe a connected manifold whose universal covering space is contractible. Thus quite simply we are concerned with closed, connected manifolds which are $K(\pi,1)$-spaces. Our original proof of (30.1) made use of heavy computation with characteristic numbers, however Royster's argument [Ro_1] makes a simple discussion possible, illustrating how various of the previously introduced techniques can be woven together to obtain the theorem. The closed aspherical manifolds appear in differential geometry, Lie group theory, automorphic function theory and are quite common in three dimensional topology. Among closed 2-manifolds only the 2-sphere and $RP(2)$ fail to be aspherical.

If (T,M^n) is an involution on a closed aspherical manifold then in the process of defining $\Gamma[T,M^n]_2$ we introduced $V^{n+1} = (S^1 \times M^n)/(A \times T)$ which fibres over $RP(1)$ with fibre M^n and structure group C_2. It is an elementary exercise with the homotopy exact sequence of a fibration to see that if M^n were a closed aspherical manifold then so is V^{n+1}. Thus iteration of the Γ-construction will be our basic technique for building aspherical manifolds.

Let us begin with the even dimensional generators, which are by far the easier. On $RP(2)$ there is the involution $\tau[x_1,x_2,x_3] = [-x_1,x_2,x_3]$.

(30.2) LEMMA: *For any* $k \geq o$ *the bordism class* $\varepsilon(\Gamma^{2k}[\tau,RP(2)]_2) \in MO_{2(k+1)}$ *is indecomposable and may be represented by a closed aspherical manifold.*

PROOF: First we dispose of indecomposability. The fixed set of $(\tau,RP(2))$ is the disjoint union of a point and an $RP(1)$. Thus we have $\eta \to F$. Referring back to (25.6) we can say

$$\varepsilon(\Gamma^{2k}[\tau,RP(2)]) = [RP(\eta \oplus (2k+1)R)]_2 + \textstyle\sum_o^{2k-1}(\varepsilon\Gamma^j[\tau,RP(2)]_2)[RP(2k-j)]_2.$$

Obviously we want to show $[RP(\eta \oplus (2k+1)R]_2$ is indecomposable. But this is the disjoint union of $RP(2(k+1))$ and $RP(\xi_2 \oplus (2k+1)R)$ where $\xi_2 \to RP(1)$ is the line bundle associated with (A,S^1). By (21.8), $[RP(\xi_2 \oplus (2k+1)R)]_2 = o$ and we know $[RP(2(k+1))]_2$ is indecomposable.

It remains now to exhibit an involution (T,K^2) over a closed aspherical surface for which $[T,K^2]_2 = [\tau,RP(2)]_2 \in I_2(C_2)$. It is perhaps surprising that any non-trivial involution on a closed aspherical manifold of odd Euler characteristic will do. Only observe that by (28.2) such an involution would have an odd number of isolated fixed points while by (28.3) the fixed set will also contain an odd number of copies of

$\xi_2 \to RP(1)$. We can safely leave it to the reader to apply (25.1) and draw the obvious conclusion that $[T,K^2]_2 = [\tau,RP(2)]_2$ in $I_2(C_2)$. As an example think of the 2-torus T^2 as $S^1 \times S^1$. Then the involution on T^2 given by $(z_1,z_2) \to (\bar{z}_1,\bar{z}_2)$ has exactly 4 fixed points. Using one of the fixed points and the isolated fixed point in $(\tau,RP(2))$ form the connected sum $T^2 \# RP(2) = K^2$. There results an involution on the closed aspherical surface having 3 isolated fixed points and one copy of $\xi_2 \to RP(1)$ as fixed set. In any case we do have even dimensional aspherical generators. ∎

Next we shall introduce a bilinear pairing

$$MO_*(C_2) \otimes_{MO_*} I_*(C_2) \to MO_*.$$

To denote an element in $I_n(C_2)$ we preserve the notation $[T,M^n]_2$, however if (τ,V^m) is fixed point free we shall use $\{\tau,V^m\}_2 \in MO_m(C_2)$. Let us define $<\{\tau,V^m\}_2,[T,M^n]_2>$ in MO_{n+m} to be the unoriented bordism class of the quotient $(V^m \times M^n)/(\tau \times T)$. This quotient fibres over V^m/T with fibre M^n and structure group C_2. These statements now follow

i) *if* $y = [T,M^n]_2 \in I_n(C_2)$ *then* $<\{A,S^0\}_2,y> = \varepsilon(y) = [M^n]_2$

ii) *if* $x = \{\tau,V^m\}_2 \in MO_m(C_2)$ *then* $<x,1> = \varepsilon_*(x) = [V^m/\tau]_2$

iii) *for any* $x \in MO_*(C_2)$, $y \in I_*(C_2)$, $<\gamma(x),y> = <x,\Gamma(y)>$.

We comment on the last remark which involves γ from Section 24 and Γ from Section 25. Think of it as follows. On $V^m \times S^1 \times M^n$ introduce two commuting involutions

$$\alpha_1(x,z,y) = (\tau x,\bar{z},y)$$

$$\alpha_2(x,z,y) = (x,-z,Ty).$$

If the involution α_1 is first factored out we have $\gamma(\tau,V^m) \times M^n$ and if the involution induced by α_2 on this product is now factored out we have $<\gamma(\tau,V^m), (T,M^n)>$. On the other hand if we first form the quotient by α_2 we have $V^m \times \Gamma(T,M^n)$ and then by factoring out the involution on this product induced by α_1 we receive $<(T,V^m),\Gamma(T,M^n>$. Since α_1, α_2 commute the equality in iii) follows. This tells us that γ and Γ are adjoints with respect to this pairing.

In particular $\varepsilon\Gamma^k[T,M^n]_2 = <\{A,S^0\}_2,\Gamma^k[T,M^n]_2> = <\gamma^k\{A,S^0\}_2,[T,M^n]_2> = <\{T,X(k)\}_2,[T,M^n]_2>$ where $\{T,X(k)\}_2$ is the element of the natural basis of $MO_*(C_2)$ introduced in Section 24.

An element of $I_n(C_2)$ is said to be *decomposable* if and only if it can be expressed as a sum of products of lower dimensional elements.

(30.3) LEMMA: *If* $x \in I_n(C_2)$ *is decomposable then so is* $\varepsilon\Gamma^k(x) \in MO_{n+k}$ *for all* k. *For any* $x \in I_n(C_2)$

$$<\{A,S^k\}_2,x> = \varepsilon\Gamma^k(x)$$

modulo a decomposable.

PROOF: By (25.7), Γ preserves the ideal of decomposables and certainly ε does too. For the second assertion

$$\{A,S^k\}_2 = \{T,X(k)\}_2 + \textstyle\sum_1^k\{T,X(k-j)\}_2[RP(j)]_2$$

so that

$$<\{A,S^k\}_2,x> = <\{T,X(k)\}_2,x \quad + \textstyle\sum_1^k<\{T,X(k-j)\}_2,x>[RP(j)]_2$$

and

$$\varepsilon\Gamma^k(x) = <\{T,X(k)\}_2,x>. \quad ■$$

To use these results we need $D : MO_n \to I_{2n}(C_2)$. We send $[M^n]_2 \to [d,M^n \times M^n]_2$ where $d(x,y) = (y,x)$. Using (25.1), since the normal bundle to the fixed point set agrees with the tangent bundle to M^n this is well defined. Even more, $D : MO_* \to I_*(C_2)$ is a ring homomorphism which doubles dimension. This will also follow from (25.1), but it can be seen geometrically also. The multiplicative part is easy and the additive part will use that a fixed point free involution on a closed manifold is zero in $I_*(C_2)$. Try it. All we have done is to factor the squaring endomorphism of MO_* through $MO_* \xrightarrow{D} I_*(C_2) \xrightarrow{\varepsilon} MO_*$.

Suppose we can write $n = 2^{k+1}m + 2^k - 1$ with $m > 0$, $k > 0$.

(30.4) LEMMA: *If* $V^{2^k m}$ *is a closed aspherical manifold with indecomposable bordism class, then* $<(T,X(2^k-1)),(d,V \times V)>$ *is also a closed aspherical manifold of dimension* $n = 2^{k+1}m + 2^k - 1$ *with an indecomposable bordism class.*

PROOF: The first step is to see that $<\{A,S^{2^k-1}\}_2,[d,RP(2^km) \times RP(2^km)]_2>$ is indecomposable. Let $(c,CP(2^km))$ be the conjugation involution, then our discussion of conjugations in Section 22 proves, in view of (25.1), that $[c,CP(2^km)]_2 = [d,RP(2^km) \times RP(2^km)]_2$ in $I_{2^{k+1}m}(C_2)$. Now the manifold $<\{A,S^{2^k-1}\}_2,[c,CP(2^km)]_2>$ is exactly the indecomposable generator constructed by Dold in [Do_2].

Since $x = [RP(2^k m)]_2$ and $y = [V^{2^k m}]_2$ are both indecomposable the sum $x+y$ is decomposable. Thus so is $D(x+y)$ and hence by (30.3), $\varepsilon\Gamma^{2^k-1}D(x) = \varepsilon\Gamma^{2^k-1}D(y)$ modulo a decomposable also. We just showed that $<\{A,S^{2^k-1}\}_2, D(x)>$ is indecomposable. This completes the proof of (30.1) because we had already constructed closed aspherical manifolds in all even dimensions with indecomposable bordism classes. ∎

This result suggests the possibility that every unoriented bordism class contains a closed connected aspherical representative. The difficulty lies in the fact that the connected sum operation in dimensions larger than 2 destroys asphericity.

31. Actions of $(C_2)^k$ without stationary points

We denote by $(C_2)^k$ the elementary abelian 2-group formed by taking the k-fold direct product of C_2 with itself. An action of $(C_2)^k$ is equivalent to a sequence of pair-wise commuting involutions $T_j : M^n \to M^n$, $j = 1,\dots,k$. A *stationary point* of the action is a point fixed under all the T_j, and hence fixed by all group elements in $(C_2)^k$. We are concerned with smooth actions of $(C_2)^k$ on closed manifolds. In this section we shall show

(31.1) THEOREM: *If* $(C_2)^k$ *acts smoothly on a closed manifold* M^n *without stationary points then* $[M^n]_2 = o$.

PROOF: Certainly we have long known this for $k = 1$. Inductively we assume the result for $(C_2)^{k-1}$. Consider an action $((C_2)^k, M^n)$ with no stationary points. We write $(C_2)^k = C_2 \times (C_2)^{k-1}$. Let $F \subset M^n$ denote the fixed point set of the first C_2. If $F = \emptyset$ then we are done by the inductive assumption. Suppose $F \neq \emptyset$, then $C_2 \times (C_2)^{k-1}$ acts as a group of orthogonal bundle maps on the normal bundle $\eta \to F$. The generator, T, of the first C_2 acts on each fibre as the antipodal map, however, as $(C_2)^{k-1}$ has no stationary point in F there is no fibre of $\eta \to F$ carried into itself by all the elements of $(C_2)^{k-1}$. We form the Whitney sum $\eta \oplus R \to F$ and, with (v,t) in $\eta \oplus R$, we extend the action of $C_2 \times (C_2)^{k-1}$ to $\eta \oplus R$ by $T(v,t) = (-v,-t)$ and for $g \in (C_2)^{k-1}$, $g(v,t) = (gv,t)$. This we can pass over to the sphere bundle to give $((C_2)^k, S(\eta \oplus R))$ with T acting as the bundle involution. Certainly then there is induced an action $((C_2)^{k-1}, RP(\eta \oplus R)) \to ((C_2)^{k-1}, F)$. Since $(C_2)^{k-1}$ on F has no stationary points the action $((C_2)^{k-1}, RP(\eta \oplus R))$ is also stationary point free. Thus by the inductive hypothesis $[RP(\eta \oplus R)]_2 = o$, while from (22.2), $[M^n]_2 = [RP(\eta \oplus R)]_2 = o$. ∎

A stronger version of this result was demonstrated by Stong [St_4].

(31.2) STONG: *If* $((C_2)^k, M^n)$ *is a stationary point free action on a closed manifold then in* $I_n((C_2)^k), [((C_2)^k, M^n]_2 = 0$.

32. Actions of $C_2 \times C_2$ with isolated stationary points

If C_2 acts on a closed positive dimensional manifold with only isolated fixed points then there are an even number of fixed points and the manifold bounds. The situation for actions of $(C_2)^k$ with finitely many stationary points is much more complicated. For example $(C_2 \times C_2, RP(2))$ can be defined by $T_1[z_1,z_2,z_3] = [-z_1,z_2,z_3]$; $T_2[z_1,z_2,z_3] = [z_1,-z_2,z_s$ and it may be immediately verified that there are exactly three stationary points, $[1,0,0],[0,1,0]$ and $[0,0,1]$.

We can set up a general question as follows. Consider all finite dimensional real linear representations of the finite group G. Two such are equivalent representations if and only if they are joined by a G-equivariant real linear isomorphism of the vector spaces. We throw in a 0-dimensional representation. These equivalence classes are called *representation classes*. A positive dimensional class is *irreducible* if and only if every representation which belongs to it is irreducible. Regarding $Z/2Z$ as the field with two elements, we denote by $R_n(G)$ the vector space over $Z/2Z$ generated by the representation classes of degree n. If $R(G) = \sum R_n(G)$ then we can convert $R(G)$ into a graded commutative algebra with unit over $Z/2Z$. The product is given as follows. Suppose (G,V_1), (G,V_2) are representations. We take $(G, V_1 \oplus V_2)$ to be $g(v_1,v_2) = (gv_1,gv_2)$. Then the product is $(G,V_1) \cdot (G,V_2) = (G, V_1 \oplus V_2)$. The identity element is the representation class of degree 0. In fact $R(G)$ is the graded polynomial ring over $Z/2Z$ generated by the irreducible representation classes. We call $R(G)$ the *unoriented representation algebra*. We caution the reader against confusing this with the Grothendieck ring of representations, an entirely different concept.

Consider now an action (G,M^n) with a finite set of stationary points $x_1,\ldots,x_k$. At each stationary point, x_j, we receive a real linear representation of G on the tangent space to M^n at x_j. Denote the resulting representation class by $X(x_j) \in R_n(G)$. To (G,M^n) assign the element $X(x_1) + \ldots + X(x_k)$ in $R_n(G)$. This is $0 \in R_n(G)$ if and only if each local representation which occurs is present at an even number of the stationary points. Denote by $S_n(G) \subset R_n(G)$ the set of all such $\sum X(x_j)$ arising from all actions of G on closed n-manifolds with a finite set of stationary points. Using the disjoint union we see that $S_n(G)$ is a subspace. Furthermore, $S_*(G) = \sum S_n(G)$ is a graded subalgebra. For if G acts on M^n with stationary points $x_1,\ldots,x_k$ and on V^n with stationary points $y_1,\ldots,y_j$ then G acts on $M^n \times V^m$ by the diagonal action with stationary points (x_p,y_q). Furthermore, $X(x_p,y_q) = X(x_p)X(y_q)$ so that $\sum X(x_p,y_q) = (\sum X(x_p))(\sum X(y_q))$. Since G can act on a point, $S_*(G)$ contains the identity. In fact for $G = C_2$, $S_*(C_2)$ consists of the identity element and zero.

Suppose $I_*(G)$ is the unrestricted mod 2 bordism algebra of all actions of G on closed manifolds. Then let $Z_*(G) \subset I_*(G)$ be those bordism classes which admit a representative having isolated (possibly no) fixed (stationary) points. Then $Z_*(G)$ is a subalgebra containing the identity. The correspondence $(G,M^n) \to \sum X(x_j)$ is then easily seen to induce an algebra homomorphism $Z_*(G) \to R_*(G)$ with image $S_*(G)$. Actually Stong's theorem, (31.1), can be used to show that $Z_*((C_2)^k) \simeq S_*((C_2)^k)$.

We shall point out why this happens. Consider a representation (G,R^n) in which no non-trivial vector is stationary. Forming the one-point compactification of R^n we receive (G,S^n) an action with just two stationary points o and ∞. Clearly $X(o) = X(\infty)$ is the class of (G,R^n). Next consider an action with isolated fixed points (G,M^n). Suppose $X(x_1) = X(x_2)$ then we can cancel off these two stationary points without changing $[G,M^n]_2 \in Z_n(G)$. There is the (G,S^n) with $X(o) = X(\infty) = X(x_1) = X(x_2)$. Delete small invariant open cells around x_1,x_2 in M^n and around o,∞ in S^n. Make the obvious and appropriate boundary identifications to receive (G,V^n) with isolated stationary points, $x_3,\ldots,x_k$. The reader can show $[G,M^n]_2 = [G,V^n]_2$ because (G,S^n) extends to (G,D^{n+1}) with an arc of stationary points.

(32.1) LEMMA: *If (G,M^n) is an action with isolated stationary points then (G,M^n) is bordant to an action with isolated stationary points in which all the local representation classes are distinct.* ∎

In particular, if $\sum X(x_j) = o$ then (G,M^n) is bordant to an action without stationary points.

Let us get on with the computation of $S_*(C_2 \times C_2)$. There are four irreducible representation classes of degree 1. On the real line these are given by

$Y_0 : T_1(s) = s,\ T_2(s) = s$

$Y_1 : T_1(s) = -s,\ T_2(s) = s$

$Y_2 : T_1(s) = s,\ T_2(s) = -s$

$Y_3 : T_1(s) = -s,\ T_2(s) = -s.$

Thus R $(C_2 \times C_2)$ is the polynomial algebra $Z/2Z[Y_0, Y_1, Y_2, Y_2]$.

Suppose $C_2 \times C_2$ acts smoothly on M^n and $x \in M^n$ is an isolated stationary point. Then $X(x) = Y_1^p Y_2^p Y_3^r$ where p,q and r are non-negative integers, $p+q+r = n$, with the following significance. First p is the dimension of the component of the fixed set of T_2 containing x, the q is the dimension of the component of the fixed set of T_1 containing x and r is the dimension of the component of the fixed set of $T_1 T_2$ containing x.

As an example consider $(C_2 \times C_2, RP(2))$ mentioned earlier. With $x_1 = [1,o,o]$; $x_2 = [o,1,o]$; $x_3 = [o,o,1]$ we find $X(x_1) = Y_1 \cdot Y_3$; $X(x_2) = Y_2 \cdot Y_3$ and $X(x_3) = Y_1 \cdot Y_2$. To see the case of x_3 we can use local coordinates $[x,y,1]$ so $T_1[x,y,1] = [-x,y,1]$ and $T_2[x,y,1] = [x,-y,1$ and hence $X(x_3) = Y_1 \cdot Y_2$. We now know that $Y_1 \cdot Y_2 + Y_1 \cdot Y_3 + Y_2 \cdot Y_3$ belongs to $S_*(C_2 \times C_2)$.

(32.2) THEOREM: *The algebra* $S_*(C_2 \times C_2)$ *is the polynomial subalgebra of* $S_*(C_2 \times C_2)$ *generated by* $Y_1 \cdot Y_2 + Y_1 \cdot Y_3 + Y_2 \cdot Y_3$.

PROOF: We have just seen that the polynomial subalgebra generated by $Y_1 \cdot Y_2 + Y_1 \cdot Y_3 + Y_2 \cdot Y_3$ does belong to $S_*(C_2 \times C_2)$. We consider now, with $n > o$, a smooth action $(C_2 \times C_2, M^n)$ with a non-empty but finite collection of stationary points. In view of (32.1) we can assume that all the local representation classes are distinct.

Let us first agree that $X(x_i) = Y_1^n$ does not occur as a local representation. Suppose it did, then T_2 leaves every point fixed in an open invariant neighborhood of x_i and therefore T_2 leaves the whole component, V^n, of M^n containing x_i pointwise fixed. Now T_1 must have at least two fixed points in V^n so there is another stationary point x_j with $X(x_j) = Y_1^n$ which would contradict the uniqueness of the local representations. Obviously this proves Y_2^n and Y_3^n cannot occur either.

With $X(x_j) = Y_1^{p_j} Y_2^{q_j} Y_3^{r_j}$ we consider the sequence of numbers $p_1, q_1, r_1, p_2, q_2, r_2, \ldots$. To be definite we will suppose p_1 is the largest integer occuring in this sequence. Of all j with $p_j = p_1$ let us assume for convenience that

$r_1 = \max\{r_j | p_j = p_1\}$.

(We might need only to renumber). Then if $p_j = p_1$ for $j \neq 1$ we must have $r_j < r_1$ for if $r_j = r_1$ we would also have $q_j = q_1$ contradicting distinctness of local representations.

We must argue that $X(x_1) = Y_1^m Y_3^m$ and $n = 2m$. We denote by $F(T_1, M^n)$ the set of fixed points of T_1. As explained earlier the dimension of the component of $F(T_1, M^n)$ passing through the stationary point x_j is given by the integer q_j. We turn to the action of $(C_2 \times C_2, S(\eta)) \to (C_2 \times C_2, F(T_1, M))$ on the normal sphere bundle to the set of points fixed under T_1. We want to analyse $F(T_2, S(\eta)) = F(T_2, M) \cap S(\eta)$. Each point of this intersection must lie in a fibre of $S(\eta)$ over a stationary point of $C_2 \times C_2$. Over the point x_j the $F(T_2, M) \cap S(\eta)$ will be the sphere S^{p_j-1}. This is contained in the fibre sphere $S^{p_j+r_j-1}$. Furthermore, T_1 acts upon S^{p_j-1} as the antipodal involution.

(32.3) LEMMA: *Any positive value of* p_j *which occurs at some stationary point will occur at an even number of stationary points. This is also true for positive values of* q_j *and* r_j.

PROOF: We shall indicate this for p_1. The compliment of the interior of the normal tube around $F(T_1, M)$ is a compact $C_2 \times C_2$ manifold B^n. Now T_1 acts freely on B^n and hence on $F(T_2, B^n) \subset B^n$. But $(T_1, \dot{F}(T_2, B^n)) = (T_1, F(T_2, S(\eta)))$ and hence $\sum_{x_j} [T_1, S^{p_j-1}]_2 = o \in MO_*(C_2)$. Splitting this up according to dimension we then find, for example,

$$\sum_{p_j = p_1} [T_1, S^{p_j-1}]_2 = o \in MO_{p_1-1}(C_2)$$

and therefore p_1 occurs at an even number of stationary points. ∎

We need a little more detailed information. Specifically what is the form of the total Whitney class of the normal bundle to $RP(p_j-1)$ in $S(\eta)/T_1 = RP(\eta)$. We claim it is $(1+d)^{r_j}$ where $d \in H^1(RP(p_j-1)); Z/2Z)$ is the generator. It is clear that this is the class of the normal bundle to $RP(p_j-1)$ in the fibre $RP(p_j+r_j-1)$. But the normal bundle to the fibre $RP(p_j+r_j-1)$ in $RP(\eta)$ is certainly trivial so we have identified the total normal class of $RP(p_j-1) \subset RP(\eta)$ as $(1+d)^{r_j}$. We use this now to prove $p_1 = r_1$. Suppose $p_1 > r_1$. Consider then $c^{p_1-r_1-1} v_{r_1} \in H^{p_1-1}(RP(P_j-1); Z/2Z)$ for all $p_j = p_1$. Using $(T_1, \dot{B}^n) = (T_1, S(\eta))$ and $(T_1, \dot{F}(T_2, B^n)) = (T_1, F(T_2, S(\eta)))$ it follows immediately that

$$\sum_{p_j=p_1} \langle d^{p_1-r_1-1} v_{r_1}, \sigma(RP(p_j-1))\rangle = o.$$

But since $r_j < r_1$ for $j \neq 1$ we see $v_{r_1} = o$ on the corresponding $RP(p_j-1)$ because the total normal class there is $(1+d)^{r_j}$. On the other hand at $RP(p_1-1), v_{r_1} = d^{r_1}$ and $d^{p_1-r_1-1} d^{r_1} = d^{p_1-1} \neq o$. This contradiction followed by assuming $r_1 < p_1$.

So we have $p_1 = r_1$ and we must show that $q_1 = o$. If $q_1 > o$ then by (32.3) we can find another stationary point x_j, $j \neq 1$, with $q_j = q_1$. By uniqueness of local representations $(p_1,q_1,p_1) \neq (p_j,q_1,r_j)$. Since p_1 is the largest integer that occurs $p_1 \geq p_j$. Since $n = 2p_1 + q_1 = p_j + r_j + q_1$ the case $p_1 = p_j$ is eliminated. But with $p_j < p_1$ we find $r_j > p_1$, again a contradiction to the maximality of p_1. Hence $q_1 = o$. That is, $X(x_1) = Y_1^m Y_3^m$ with $n = 2m$.

We can repeat the above argument reversing the roles of r_j and q_j and replacing T_2 by T_1T_2 to prove there is an x_j with $X(x_j) = Y_1^m Y_2^m$. Let us say this is x_2. Finally consider $\max\{q_j | r_j = m\}$ and call this value q_3. Use the argument again replacing T_1 with T_2 and T_2 with T_1T_2 to show $q_3 = m$, thus finding x_3 with $X(x_3) = Y_2^m Y_3^m$. So far we have

(32.4) LEMMA: *Let* $(C_2 \times C_2, M^n)$ *be an action with a non-empty finite set of stationary points for which all the local representation classes are distinct, then* $n = 2m$ *and there are stationary points* $x_1, X(x_1) = Y_1^m Y_3^m$; $x_2, X(x_2) = Y_1^m Y_2^m$; $x_3, X(x_3) = Y_2^m Y_3^m$. ∎

Consider now $(C_2 \times C_2, RP(2))$ and the diagonal action $(C_2 \times C_2, (RP(2))^m)$ If $y_1, y_2, y_3 \dots$ are the stationary points of this latter action then $\sum X(y_j) = (Y_1 \cdot Y_2 + Y_1 \cdot Y_3 + Y_2 \cdot Y_3)^m$. We can suppose $y_1 = Y_1^m \cdot Y_3^m$, $y_2 = Y_1^m \cdot Y_2^m$ and $y_3 = Y_2^m \cdot Y_3^m$. Delete the interiors of invariant cellular neighborhoods of $x_1 \in M^{2m}$ and y_1 in $RP(2)^m$ and then identify along boundaries. Do the same for x_2, y_2 and x_3, y_3. There results a closed manifold $(C_2 \times C_2, V^{2m})$ with isolated fixed points, $z_1, \dots, z_j, \dots$ but none of the types $Y_1^m Y_3^m$, $Y_1^m Y_2^m$ or $Y_2^m Y_3^m$. It follows from (32.4) that $\sum X(z_i) = o$ and hence $\sum X(x_i) = \sum X(y_i) = (Y_1 Y_2 + Y_1 Y_3 + Y_2 Y_3)^m$. This proves (32.2). ∎

(32.5) COROLLARY: *Let* $(C_2 \times C_2, M^n)$ *be an action with isolated stationary points, say* $x_1, x_2, \dots$. *Either* $[M^n]_2 = o$ *or* $n = 2m$ *and* $[M^{2m}]_2 = [RP(2)]_2^m$. *If* $[M^n]_2 = o$ *then* $\sum X(x_j) = o$, *while if* $[M^{2m}]_2 = [RP(2)]_2^m$ *then* $\sum X(x_j) = (Y_1 Y_2 + Y_1 Y_3 + Y_2 Y_3)^m$.

PROOF: Either $\sum X(x_j) = o$ or $n = 2m$ and $\sum X(x_j) = (Y_1 Y_2 + Y_1 Y_3 + Y_2 Y_3)^m$ by (32.2). In the former case we apply the cancellation procedure of (32.1) to replace $(C_2 \times C_2, M^n)$ with a stationary point free action $(C_2 \times C_2, V^n)$ for which $[C_2 \times C_2, V^n]_2 = [C_2 \times C_2, M^n]_2$. From (31.1) then $[M^n]_2 = [V^n]_2 = o$.

If $n = 2m$ and $\sum X(x_j) = (Y_1 \cdot Y_2 + Y_1 \cdot Y_3 + Y_2 \cdot Y_3)^m$ we use the connected sum of $(C_2 \times C_2, M^n)$ and $(C_2 \times C_2, RP(2)^m)$ to produce a $(C_2 \times C_2, V^n)$ which goes to o in $S_n(C_2 \times C_2)$. Then

$$[V^n]_2 = [M^n]_2 + [RP(2)]_2^m = o. \quad \blacksquare$$

The following actions of $C_2 \times C_2$ might be noted. It acts on the complex and quaternionic projective planes just as on RP(2) with three stationary points in each case. This agrees with the known facts that $[CP(2)]_2 = [RP(2)]_2^2$ and $[QP(2)]_2 = [RP(2)]_2^4$. There is even a similar action on the Caley plane which is bordant to the 8 power of RP(2).

Let us round off the section with

(32.6) THEOREM: *If* $(C_2)^k$ *acts smoothly on a closed manifold,* $n > o$, *then there cannot be exactly one stationary point.*

PROOF: Suppose $x \in M^n$ is a stationary point, then we receive a representation of degree n in the tangent space of M^n at x. This is a sum of non-trivial representations of degree 1 so we can find a subgroup $(C_2)^{k-1}$ for which the subspace of fixed vectors has positive dimension. Accordingly, if $F \subset M^n$ is the set of stationary points of $((C_2)^{k-1}, M^n)$ then the component, C, of F containing x has positive dimension and is $(C_2)^k$-invariant. We can, by an appropriate choice of generator, T, write $(C_2)^k = C_2 \times (C_2)^{k-1}$. Then $T : C \to C$ has a fixed point x and since C is connected and positive dimensional T has another fixed point $y \in C$. Clearly y is also a stationary point of $((C_2)^k, M^n)$. $\blacksquare$

Results of this sort do not hold for actions of C_4; that is, for maps of period 4. For example C_4 can act on RP(2k) with precisely one stationary point. Describe the closed 2k-disk in complex coordinates as

$$D^{2k} = \{(z_1,\dots,z_k) \mid \sum z_j \bar{z}_j \leq 1\}.$$

On this disk we act with C_4 by introducing $T(z_1,\dots,z_k) = (iz_1,\dots,iz_k)$, $i = \sqrt{-1}$. If antipodal points on $\dot{D}^{2k} = S^{2k-1}$ are identified then there is induced on RP(2k) a diffeomorphism of period 4 with only one stationary point. In particular we have actions $(C_4, RP(4))$ and $(C_4, RP(2) \times RP(2))$ each of which has only a single stationary point.

It is seen that open cellular neighborhoods of these stationary points may be deleted and the resulting manifolds identified along their spherical boundaries. There results an action (C_4, M^4) on this connected sum with no stationary points. Further $[M^4] = [RP(4)]_2 + [RP(2)]_2^2 \neq 0$, hence (31.1) does not extend to C_4 either. For additional discussion of C_{2^k} actions we refer to Stong [St_3].

33. The bundle involution

The results in the two final sections of this chapter actually do not concern bordism, but do involve several other concepts which have already been brought up.

First we need a particular device for introducing the tensor product of a line bundle with an n-plane bundle. Suppose we are given a fixed point free involution (T,X) together with an n-plane bundle $p : \eta \to X/T$ over the quotient space. The quotient map $\nu : X \to X/T$ will then pull back an n-plane bundle $\tilde{\eta} \to X$. Here $\tilde{\eta} \subset X \times \eta$ is given by $\{(x,v) \mid \nu(x) = p(v)\}$ and $\tilde{p} : \tilde{\eta} \to X$ is $(x,v) \to x$. There is a fixed point free involution $(\tilde{T},\tilde{\eta}) \to (T,X)$ given by $\tilde{T}(x,v) = (Tx,-v)$. The equivariant $\tilde{p} : (\tilde{T},\tilde{\eta}) \to (T,X)$ then induces $\hat{p} : \tilde{\eta}/\tilde{T} = \hat{\eta} \to X/T$. It is seen that this is again an n-plane bundle $\hat{\eta} \to X/T$, which we refer to as the *twist* of $\eta \to X/T$ by (T,X). Let $\xi \to X/T$ be the real line bundle associated to (T,X).

(33.1) LEMMA: *The twist* $\hat{\eta} \to X/T$ *is equivalent to the tensor product* $\xi \otimes \eta \to X/T$. *Furthermore, the total Whitney class of* $\hat{\eta}$ *is given by* $\sum_0^n (1+c)^k v_{n-k}$.

PROOF: We must define a bundle map, φ, so that commutativity holds in

$$\begin{array}{ccc} \hat{\eta} & \xrightarrow{\varphi} & \xi \otimes \eta \\ & \searrow \quad \swarrow & \\ & X/T & \end{array}$$

Consider $\tilde{\eta} = \{(x,v) \mid \nu(x) = p(v)\} \subset X \times \eta$. We identify X with the o-sphere bundle associated with $\xi \to X/T$. Thus $T(x) = -x$ and we think of $X \times \eta \subset \xi \times \eta$ to define $\Psi : \tilde{\eta} \to \xi \otimes \eta$ by $\Psi(x,v) \to x \otimes v$. Now $\Psi(\tilde{T}(x,v)) = \Psi(-x,-v) = (-x \otimes -v) = x \otimes v$ so Ψ will induce $\varphi : \tilde{\eta}/\tilde{T} \to \xi \otimes \eta$. This is the required bundle map so the first assertion follows.

The Whitney class of $\xi \otimes \eta$ is computed as follows. Write the total Whitney class of $\eta \to X/T$ in factored form $(1+t_1)\ldots(1+t_n)$. Let $c \in H^1(X/T;Z/2Z)$ be the first Whitney class of $\xi \to X/T$ (i.e. the characteristic class of the involution (T,X)). Then the total Whitney class of the tensor product $\xi \otimes \eta$ is $(1+c+t_1)\ldots(1+c+t_n) = \sum_0^n (1+c)^k v_{n-k}$. The last step recalls that v_{n-k} is the $(n-k)$-th elementary symmetric function in $t_1,\ldots,t_n$. ■ We might now give a proof of the Borel-Hirzebruch theorem (21.3). Let $p : \alpha \to X$ be an orthogonal k-plane bundle. There is the unit sphere bundle $(T,S(\alpha))$ with the bundle involution and $S(\alpha)/T = RP(\alpha) \to X$ is a bundle with fibre $RP(k-1)$ and structure group $PO(k)$. We consider the tangent bundle along the fibres $\eta \to RP(\alpha)$, which is a $(k-1)$-plane bundle. There is the bundle $\eta_1 \to S(\alpha)$ where $\eta_1 \subset S(\alpha) \times \alpha$ consists of all pairs (v_1,v_2) with $p(v_1) = p(v_2)$ and v_2 orthogonal to v_1. The projection is $(v_1,v_2) \to v_1$. This is the bundle of tangent vectors along the fibres in $S(\alpha) \to X$. Moreover, $(T,S(\alpha))$ induces an involution $(T',n_1) \to (T,S(\alpha))$ given by $T'(v_1,v_2) = (-v_1,-v_2)$. The quotient bundle, $\eta_1/T' \to S(\alpha)/T = RP(\alpha)$, is the tangent bundle along the fibres $\eta \to RP(\alpha)$.

Now from the diagram

$$\begin{array}{ccc} & & \alpha \\ & & \downarrow \\ RP(\alpha) & \longrightarrow & X \end{array}$$

there is pulled back a k-plane bundle $\beta \to RP(\alpha)$. We must show that the twist of β by $(T,S(\alpha))$ is $R \oplus \eta \to RP(\alpha)$. We simply note $\beta_1 = R \oplus \eta_1 \to S(\alpha)$ in the following natural way. Think of $\beta_1 \subset S(\alpha) \times \alpha$ as all pairs (v_1,v) with $p(v_1) = p(v)$. Then v can be uniquely expressed as $v = rv_1 + v_2$ for a real number r and a vector v_2 which is orthogonal to v_1. This shows $\beta_1 = R \oplus \eta_1$. In forming the twist we shall again find $\hat{\beta}$ also splits as a Whitney sum. According to the preceding remarks the factor contributed by η_1 is just $\eta \to RP(\alpha)$. On the other hand the trivial factor contributes the twist of the canonical real line bundle $\xi \to RP(\alpha)$, which is again trivial. Hence $\hat{\beta} = R \oplus \eta$ so that $\hat{\beta}$ and η have the same Whitney class. When the total Whitney class of $\hat{\beta}$ is computed from (33.1) the Borel-Hirzebruch formula then follows.

In this section we shall make use of the following application of the twisting operation.

(33.2) LEMMA: *Let* (T, M^k) *be a fixed point free involution on a closed manifold and let* (T', V^{n+m}) *be an involution on a closed manifold for which* F^m *is non-empty. Let* $c \in H^1(M^k/T; Z/2Z)$ *be the fundamental class of* (T, M^k) *and let* $\eta \to F^m$ *be the normal bundle to* $F^m \subset V^{n+m}$*. The total Whitney class of the normal* n*-plane bundle to* $(M^k/T) \times F^m$ in $(M^k \times V^{n+m})/(T \times T')$ *is given by* $\sum_0^n (1 \oplus c)^j \otimes v_{n-j}(\eta)$.

PROOF: According to Section 20 we identify $D(\eta)$ with a closed invariant tubular neighborhood N of F^m in V^{n+m}. Furthermore, the antipodal map, A, of $D(\eta)$ is identified with $T' : N \to N$. Now $(M^k \times N)/(T \times T')$ is a tubular neighborhood of $(M^k \times F^m)/(T \times T') = M^k/T \times F^m$ in $(M^k \times V^{n+m})/(T \times T')$. But $(M^k \times N)/(T \times T')$ identifies with $(M^n \times D(\eta))/(T \times A)$. Thus it is seen that the normal bundle to $(M^k/T) \times F^m$ is obtained from the twist by the involution $(T \times id, M^k \times F^m)$ with the bundle induced by the projection

$$\begin{array}{ccc} & & \eta \\ & & \downarrow \\ (M^k/T) \times F^m & \to & F^m \end{array}$$

The total Whitney class follows from (33.1). ∎

To apply this we shall make use of the Smith-Gysin exact sequence [B_4] about which we now comment briefly. Let (T, X) be a fixed point free involution of a closed manifold. There is then the quotient map $\nu : X \to X/T$, the cohomology class $c \in H^1(X/T; Z/2Z)$ and a Gysin homomorphism $\nu_* : H^j(X; Z/2Z) \to H^j(X/T; Z/2Z)$ described as follows. There are mod 2 fundamental homology classes $\sigma(X)$ and $\sigma(X/T)$. For $h \in H^j(X; Z/2Z)$ we define $\nu_*(h) \in H^j(X/T; Z/2Z)$ to be the unique cohomology class for which $\nu_*(h \cap \sigma(X)) = \nu_*(h) \cap \sigma(X/T))$. There is then an exact cohomology sequence

$$\ldots \xrightarrow{\nu_*} H^j(X/T; Z/2Z) \xrightarrow{c} H^{j+1}(X/T; Z/2Z) \xrightarrow{\nu^*} H^{j+1}(X; Z/2Z)$$
$$\xrightarrow{\nu_*} H^{j+1}(X/T; Z/2Z) \longrightarrow \ldots$$

Here we have used c to denote cup-product with the characterstic cohomology class $c \in H^1(X/T; Z/2Z)$. This is simply the Gysin sequence of a o-sphere bundle.

(33.3) THEOREM: *Let* (T, V^{n+m}) *be an involution on a closed manifold for which* F^m *is non-empty. If* $H^j(V^{n+m}; Z/2Z) = o$, $n - k \leq j \leq n$ *then for the normal bundle to* F^m *in* V^{n+m} *we have* $v_j = o$ *for* $n - k \leq j \leq n$.

PROOF: Select an $r > n$ and consider the fixed point free involutions $(T' = A \times T,\ S^r \times V^{m+n})$ and $(A \times \mathrm{id},\ S^r \times F^m)$. We shall write $d \in H^1(RP(r) \times F^m; Z/2Z)$ for $d \otimes 1$, the fundamental class of the latter involution.

We shall be concerned with the (obviously) commutative diagram

$$\begin{array}{ccc} H^{n-k-1}((S^r \times V^{n+m})/T') & \xrightarrow{\ \cup\, c^{k+1}\ } & H^n((S^r \times V^{n+m})/T') \\ \Big\downarrow i^* & & \Big\downarrow i^* \\ H^{n-k-1}(RP(r) \times F^m) & \xrightarrow{\ \cup\, d^{k+1}\ } & H^n(RP(r) \times F^m). \end{array}$$

Since $H^j(V^{n+m}; Z/2Z) = 0$ for $n - k \leq j \leq n$, it follows that with $r > n$, $H^j(S^r \times V^{n+m}; Z/2Z) = 0$ for $n - k \leq j \leq n$ also. From the Smith-Gysin sequence we therefore conclude

$$\cup\, c^{k+1} : H^{n-k-1}((S^r \times V^{n+m})/T') \to H^n((S^r \times V^{n+m})/T')$$

is an epimorphism.

Now there is $\varphi_n \in H^n((S^r \times V^{n+m})/T')$ which is the cohomology class dual to the submanifold $RP(r) \times F^m \subset (S^r \times V^{n+m})/T'$. It has been shown generally by Thom [T_1] that $i^*(\varphi_n)$ would be the n-th Whitney class of the normal bundle to $RP(r) \times F^m$. Therefore by (33.2)

$$i^*(\varphi_n) = d^n \otimes 1 + d^{n-1} \otimes v_1 + \ldots + 1 \otimes v_n.$$

However, by our observation that $\cup\, c^{k+1}$ is an epimorphism in the commutative diagram we can write $\varphi_n = c^{k+1} \cup \gamma$ and

$$\sum_0^n d^i \otimes v_{n-i} = d^{k+1} i^*(\gamma).$$

But actually $i^*(\gamma) = d^{n-k-1} \otimes \gamma_0 + \ldots + 1 \otimes \gamma_{n-k-1}$ so that $\sum_0^n d^i \otimes v_{n-i} = \sum_{k+1}^n d^i \otimes \gamma_{n-i}$. We can only conclude that $v_j = 0$, $n - k \leq j \leq n$. ∎

For an alternative approach to this result see [Br_1, IV.10].

34. The Borsuk antipode theorems

In the following M^k will denote a smooth manifold, not necessarily closed or even compact. For any map $f : S^n \to M^k$ let $A(f) \subset S^n$ denote the set of all points x with $f(x) = f(-x)$. In this section we are going to prove

(34.1) THEOREM: *If* $f : S^n \to M^k$ *is a map with* $n > k$ *then dim* $A(f) \geq n - k$. *If* $f : S^n \to M^n$ *has* $f^* : H^n(M^n;Z/2Z) \to H^n(S^n;Z/2Z)$ *trivial, then* $A(f) \neq \emptyset$.

For the case M^k a closed manifold the argument we present here, which is from the first edition, is really not an unreasonable one in view of all the preceding discussions. However, in passing to the non-closed case it would not be surprising if the reader began to wonder if this is not all rather an improbable approach in which the assumption of smoothness on M^k is unnecessary. Indeed Munkholm [Mu] has made extensive generalizations of (34.1) which remove the differentiability hypothesis. The classical Borsuk-Ulam mapping theorem asserts that for $f : S^n \to R^n$, $A(f) \neq \emptyset$. We are also generalizing Borsuk's theorem to the effect that if $f : (A,S^n) \to (A,S^n)$ is equivariant then f has odd degree. The result (34.1) has some overlap with results of Bourgin [Bu] and Yang [Y_1,Y_2] concerning maps of S^n into R^k.

PROOF: We begin with the assumption that M^k is closed and connected. There is then the fixed point free involution $(T, S^n \times M^k \times M^k)$ given by $T(x,y,z) = (-x,z,y)$, and $S^n \times M^k \times M^k/T$ is a closed $(n+2k)$-manifold. The projection $S^n \times M^k \times M^k \to S^n$ induces the bundle map $p : S^n \times M^k \times M^k/T \to RP(n)$, with fibre $M^k \times M^k$ and structure group C_2. There is the invariant subset $S^n \times \Delta \subset S^n \times M^k \times M^k$ where Δ is the diagonal of $M^k \times M^k$ and hence $RP(n) \times \Delta$ is a closed $(n+k)$-dimensional submanifold in $S^n \times M^k \times M^k/T$. By (33.2) then the k-dimensional characteristic class of the normal bundle to $RP(n) \times \Delta$ is

$$v_k = \sum_o^k d^j \otimes w_{k-j}$$

where w_{k-j} is a Stiefel-Whitney class of the tangent bundle to M^k.

Put now $X = S^n \times M^k \times M^k$, then $\varphi_k \in H^k(X/T;Z/2Z)$ is dual to the submanifold $i : RP(n) \times \Delta \subset X/T$ while $i^*(\varphi_k) = \sum_o^k d^j \otimes w_{k-j}$.

If N is a closed tubular neighborhood of $RP(n) \times \Delta$ in X/T and $\sigma \in H_{n+k}(RP(n) \times \Delta;Z/2Z)$ is the fundamental class then φ_k is the image of σ under the composition

$$H_{n+k}(RP(n) \times \Delta;Z/2Z) \simeq H_{n+k}(N;Z/2Z) \simeq H^k(N,\dot{N};Z/2Z) \simeq$$
$$\simeq H^k(X/T,X/T \setminus N^o;Z/2Z) \to H^k(X/T;Z/2Z).$$

It then follows that for any open set $U \supset RP(n) \times \Delta$, φ_k is in the kernel of the homomorphism $H^k(X/T;Z/2Z) \to H^k(X/T \setminus U;Z/2Z)$.

We are now ready to consider maps $f : S^n \to M^k$. To each such map we associate a cross-section of $p : X/T \to RP(n)$ by $s((x)) = ((x,f(x),f(-x)))$ where $(x) \in RP(n)$ corresponds to $x \in S^n$ and $((x,f(x),f(-x)))$ corresponds to $(x,f(x),f(-x))$ in X. Note $((-x,f(-x),f(x)) = ((x,f(x),f(-x)))$ so that s is well defined. If $f_t : S^n \to M^k$ is a homotopy of f, then the induced homotopy of the section s is $s_t((x)) = ((x,f_t(x),f_t(-x)))$. To each $f : S^n \to M^k$ we associate the cohomology class $S^*(\varphi_k)$ in $H^k(RP(n);Z/2Z)$.

(34.2) LEMMA: *If* $s^*(\varphi_k) \neq o$, *then* $\dim A(f) \geq n - k$.

PROOF: There is the quotient map $\nu : S^n \to RP(n)$ and we put $B(f) = \nu(A(f))$. Note that $s : RP(n) \to X/T$ has $s^{-1}(RP(n) \times \Delta) = B(f)$. Let U be an open neighborhood of $RP(n) \times \Delta$ in X/T. Consider the diagram

$$\begin{array}{ccc} H^k(X/T) & \xrightarrow{j^*} & H^k(X/T \setminus U) \\ \downarrow s^* & & \downarrow s_1^* \\ H^k(RP(n)) & \xrightarrow{j_1^*} & H^k(RP(n) \setminus s^{-1}(U)). \end{array}$$

Since $j^*(\varphi_k) = o$ then $j_1^*(s^*(\varphi_k)) = j_1^*(d^k) = o$. Given an open neighborhood of B(f), call it V, we can always choose U so that $s^{-1}(U) \subset V$. Hence for every open neighborhood of B(f), d^k lies in the kernel of $H^k(RP(n)) \to H^k(RP(n) \setminus V)$. In terms of Alexander-Wallace-Spanier cohomology the support of d^k lies in B(f). Suppose now that $H^{n-k}(RP(n)) \to H^{n-k}(B(f))$ is trivial. Since we are dealing with a continuous cohomology theory we can choose an open neighborhood $V \supset B(f)$ so that d^{n-k} lies in the kernel of $H^{n-k}(RP(n)) \to H^{n-k}(\bar{V})$. Now choose $\alpha_{n-k} \in H^{n-k}(RP(n),\bar{V})$ and $\beta_k \in H^k(RP(n),RP(n) \setminus V)$ with images d^{n-k} and d^k respectively under

$$H^{n-k}(RP(n),\bar{V}) \to H^{n-k}(RP(n))$$

$$H^k(RP(n),RP(n) \setminus V) \to H^k(RP(n)).$$

Using the relative cup-product we now find d^n is the image of $\alpha_{n-k}\beta_k$ under $H^n(RP(n),\bar{V} \cup (RP(n) \setminus V)) \to H^n(RP(n))$. Since $RP(n) = \bar{V} \cup (RP(n) \setminus V)$ we have arrived at the contradiction $d^n = o$. Thus $H^{n-k}(RP(n)) \to H^{n-k}(B(f)$ is non-trivial and $\dim A(f) \geq n - k$. ■

This line of reasoning is due originally to Yang $[Y_1,Y_2]$. The reader can find more details in $[CF_1]$. In the notation of that paper we have shown $\text{co-ind}_{Z/2Z}(A(f)) \geq n - k$ if $s^*(\varphi_k) \neq o$.

First we shall show that $s^*(\varphi_k) \neq o$ if $n > k$. Certainly $s^*(\varphi_k)$ depends only on the homotopy class of the original map $f : S^n \to M^k$. Thus without loss of generality we can assume f is constant on the southern hemisphere; that is, $f(E^n_-) = y_o \in M^k$. Consider $S^{n-1} \subset S^n$ as the equator; then $f(S^{n-1}) = y_o$. Therefore we have a commutative diagram

$$\begin{array}{ccc} RP(n-1) & \xrightarrow{s_1} & RP(n) \times \Delta \\ \downarrow{\scriptstyle i_1} & & \downarrow{\scriptstyle i} \\ RP(n) & \xrightarrow{s} & X/T \end{array}$$

wherein $s_1((x)) = ((x,y_o,y_o))$ for $(x) \in RP(n-1)$. But then $i_1^* s^*(\varphi_k) = s_1^* i^*(\varphi_k) =$

$$s_1^*(d^k \otimes 1 + \ldots + 1 \otimes w_k) = i_1^*(d^k) \in H^k(RP(n-1);Z/2Z).$$

Since $n > k$ it follows that $s^*(\varphi_k) \neq o$ and we can apply (34.2).

Now let us turn to a map $f : S^n \to M^n$ for which $f^* : H^n(M^n;Z/2Z) \to H^n(S^n;Z/2Z)$ is trivial. We continue to assume M^n is closed and connected and we still require $f(E^n_-) = y_o \in M^n$. Consider now the equivariant map $F : S^n \to M^n \times M^n$ given by $F(x) = (f(x), f(-x))$. Actually F maps S^n into the wedge $M^n \vee M^n = M^n \times y_o \cup y_o \times M^n$ because either x or $-x$ is always in E^n_-. The involution $(t, M^n \times M^n)$ given by $t(y,z) = (z,y)$ leaves the wedge invariant; $M^n \vee M^n/t = M^n$ and $F : S^n \to M^n \vee M^n$ is equivariant.

(34.3) LEMMA: *If $\bar{F} : RP(n) \to M^n$ is the map of quotient spaces associated with F, then $\bar{F}^* : H^n(M^n;Z/2Z) \to H^n(RP(n);Z/2Z)$ is trivial.*

PROOF: We may see this from the commutative diagram

$$\begin{array}{ccccccccc} & & & & H^n(M^n,y_o) & & & & \\ \downarrow{\scriptstyle f^* = o} & & \downarrow{\scriptstyle f^*} & & \downarrow{\scriptstyle f^*} & & \downarrow{\scriptstyle \bar{F}^*} & & \downarrow{\scriptstyle \bar{F}^*} \\ H^n(S^n) & \xleftarrow{\cong} & H^n(S^n,E^n_-) & \xrightarrow{\cong} & H^n(E^n_+,S^{n-1}) & \xleftarrow{\cong} & H(R(n),RP(n-1)) & \xrightarrow{\cong} & H^n(RP(n)) \end{array}$$

The first $f^* = o$ by assumption and hence $\bar{F}^* = o$ also. ∎

(34.4) LEMMA: *Under the composite of* $RP(n) \times (y_o,y_o) \xrightarrow{i_1} (S^n \times (M^n \vee M^n))/T \xrightarrow{i_2} X/T$ *we have* $i_1^* i_2^*(\varphi_n) = d^n \otimes 1$.

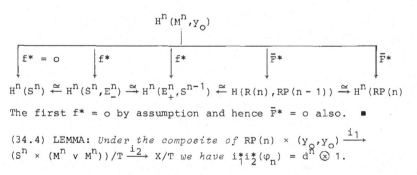

PROOF: We consider the diagram

$$\begin{array}{ccc} & \overset{i_2 i_1}{\nearrow} & X/T \\ RP(n) \times (y_o,y_o) & & \downarrow \\ & \searrow & RP(n) \end{array}$$

and recall $i^*(\varphi_n) = \sum_o^n d^{n-j} \otimes w_j \in H^n(RP(n) \times \Delta)$. ∎

Now put $Y = S^n \times (M^n \vee M^n)$ and let $s_1 : RP(n) \to Y/T$ be given by $s_1((x))$ $((x,f(x),f(-x)))$. Then $s : RP(n) \to X/T$ is the composition $i_2 s_1$. We need to show $s_1^*(i_2^*(\varphi_n)) \neq o$. Since s_1 is a cross-section of the fibre map $Y/T \to RP(n)$ it will follow that $s_1^* : H^n(Y/T;Z/2Z) \to H^n(RP(n);Z/2Z)$ is an epimorphism. Select $\gamma_n \in H^n(Y/T;Z/2Z)$ so that $s_1^*(\gamma_n) = d^n$. For example, let $\gamma_n = p_1^*(d^n)$ for $p_1 : Y/T \to RP(n)$ the fibre map. Under

$$i_1^* : H^n(Y/T;Z/2Z) \to H^n(RP(n) \times (y_o,y_o);Z/2Z)$$

we then have $i_1^*(\gamma_n) = d^n \otimes 1$. But then from (34.4) it follows that $i_1^*(\gamma_n + i_2^*(\varphi_n)) = o$ and hence the sum $\gamma_n + i_2^*(\varphi_n)$ lies in the image of

$$H^n(Y/T, RP(n) \times (y_o,y_o)) \xrightarrow{j^*} H^n(Y/T).$$

We next show that in the diagram

$$\begin{array}{ccc} H^n(Y/T, RP(n) \times y_o \times y_o) & \xrightarrow{j^*} & H^n(Y/T) \\ \simeq \big\updownarrow \beta^* & & \big\downarrow s_1^* \\ H^n(M^n, y_o) & \xrightarrow{\bar{F}^*} & H^n(RP(n)) \end{array}$$

the composition $s_1^* j^*$ is trivial.

The map

$$\beta : Y/T = (S^n \times (M^n \vee M^n))/T \to M^n \vee M^n/t = M^n$$

is induced by the projection

$$S^n \times (M^n \vee M^n) \to M^n \vee M^n.$$

It is seen that $Y/T \smallsetminus (RP(n) \times (y_o,y_o)) = S^n \times M^n \smallsetminus (S^n \times y_o)$, and that β is the projection $S^n \times (M^n,y_o) \to (M^n,y_o)$. But this projection induces an isomorphism $H^n(M^n,y_o) \simeq H^n(S^n \times (M^n,y_o))$ and hence β^* is an isomorphism in the diagram. Since $\bar{F}^* = o$ by (34.3) it follows that $s_1^* j^* = o$.

Now $\gamma_n + i_2^*(\varphi_n)$ lies in the image of j^* and thus $s_1^*(\gamma_n + i_2^*(\varphi_n)) = o$. Therefore $s^*(\varphi_n) = s_1^* i_2^*(\varphi_n) = s_1^*(\gamma_n) = d^n \otimes 1 \neq o$. This establishes (34.1) for all closed M^k.

The extension of the theorem to compact manifolds with boundary is seen immediately by doubling. An open manifold can be regarded as the increasing union of compact manifolds with boundary and since S^n is compact we also have (34.1) for open manifolds. ∎

(34.5) COROLLARY: *If* f *is a map of* S^n *into a connected non-compact* n-*manifold then there is a point* $x \in S^n$ *for which* $f(x) = f(-x)$.

PROOF: This is immediate from (34.1) since under the non-compact hypothesis $H^n(M^n;Z/2Z) = o$. We could also use here a compact connected n-manifold with non-empty boundary for the same reason. ∎

We can consider briefly what generalizations we may make for the involution (A,S^n). Suppose that (T,Σ^n) is any fixed point free involution on a closed n-manifold which is a homotopy n-sphere. There exists then equivariant maps $\varphi : (T,\Sigma^n) \to (A,S^n)$ and $\varphi' : (A,S^n) \to (T,\Sigma^n)$, [$CF_1$]. Moreover, φ' has odd degree by (34.1) as does $\varphi \circ \varphi'$ so φ is of odd degree also. Given $f : \Sigma^n \to M^k$, let $A_T(f)$ be the set of all $x \in \Sigma^n$ with $f(x) = f(Tx)$. Then obviously φ' maps $A(f \circ \varphi')$ equivariantly into $A_T(f)$. In the proof of (34.1) we really showed $\text{co-ind}_{Z/2Z}\, A(f \circ \varphi') \geq n-k$. Since an equivariant map cannot decrease co-index it follows that $\text{co-ind}_{Z/2Z}\, A_T(f) \geq n-k$ also. Thus (34.1) is also valid for (T,Σ^n).

(34.6) COROLLARY: *Any pair of fixed point free involutions* T_1,T_2 *on* S^n *have a coincidence.*

PROOF: Consider the two-to-one quotient map $\nu : S^n \to S^n/T_2$. It is immediate that $\nu^* : H^n(S^n/T_2;Z/2Z) \to H^n(S^n;Z/2Z)$ is trivial. Then from the preceding comments there is an $x \in S^n$ with $\nu(x) = \nu(T_1x)$. This means either $x = T_1x$ or $T_2x = T_1x$, but T_1 is fixed point free, so $T_2x = T_1x$. ∎

The following is a special case of a theorem of Milnor [M_1].

(34.7) COROLLARY: *If* G *is a finite group acting freely on* S^n *then every element with order* 2 *in* G *is central.*

PROOF: Suppose $g \in G$ is of order 2. For any $h \in G$, both g and hgh^{-1} give fixed point free involutions on S^n. Hence for some $x \in S^n$, $g(x) = hgh^{-1}(x)$ by (34.6). But then $g^{-1}hgh^{-1}(x) = x$ and since G acts freely it follows $g^{-1}hgh^{-1} = e \in G$ and hence g is central. ∎

We remind the reader to refer to [Mu] for further results on this subject.

CHAPTER III. MAPS OF ODD PERIOD

We have borrowed the title of this chapter from [CF_3] to describe our study of orientation preserving diffeomorphisms of odd prime power period on closed oriented manifolds. We open the chapter with a technical section aimed primarily at showing when and how a complex structure may be imposed on the normal bundle to each component of the set of fixed points. Perhaps the most obvious corollary to this is the imparting of an orientation to each of those components.

From there we pass into the determination of $MSO_*(C_p)$. This will involve a preliminary analysis of the fixed point sets.

The main objective of this chapter is dealing with the unrestricted bordism algebra $\mathcal{O}_*(C_p{}^k)$ of all orientation preserving diffeomorphisms of odd prime power period p^k. The study of $\mathcal{O}_*(C_p{}^k)$ is effected by introducing families of subgroups and considering the resulting exact sequences. A significant application will then be some of Tammo tom Dieck's results concerning the dimension of the set of stationary points in an orientation preserving action of $C_p{}^k$, [D_5]. This is an analogue in the oriented case of Boardman's work discussed in Section 27.

35. Reduction of structure group

We fix a compact Lie group H. We wish to consider right principal H-spaces (B,H) in which B is a normal Hausdorff space on which H acts freely from the right. According to a theorem of Gleason, [MZ], the quotient map $\rho : B \to B/H$ will then admit local cross-sections. We denote B/H by X and for $b \in B$, $h \in H$ we write $bh \in B$ to denote the effect of h acting on b. Let G also be a compact Lie group. We shall consider triples (G,B,H) in which (B,H) is a right principal H-space and G acts from the left on B as a group of H-equivariant homeomorphisms. That is, given $g \in G$, $h \in H$ and $b \in B$ then $g(bh) = (gb)h$. We call (G,B,H) a *right principal* H-*space with operators*. Given (G,B,H) there is induced an action of G on X by $g\rho(b) = \rho(gb)$ so that the quotient map $\rho : B \to X$ is G-equivariant.

We denote by $L \subset G$ the closed normal subgroup consisting of those elements which leave every point in X fixed. Alternatively, L is the subgroup of elements which map every H-orbit in B into itself. At each $b \in B$ we can define a homomorphism $r_b : L \to H$ as follows. Since for $g \in L$, $\rho(gb) = \rho(b)$ there is a unique element $r_b(g) \in H$ for which $gb = br_b(g)$. We note that

$$br_b(g_1g_2) = g_1g_2b = g_1(br_b(g_2)) = br_b(g_1)r_b(g_2)$$

so that $r_b(g_1g_2) = r_b(g_1)r_b(g_2)$, showing that $r_b : L \to H$ is a homomorphism. Since $h \to bh$ maps H homeomorphically onto the orbit containing b, it follows that r_b is continuous. For $h \in H$, $g \in L$ we have

$$g(bh) = (gb)h = (br_b(g))h = bh(h^{-1}r_b(g)h)$$

and thus $r_{bh}(g) = h^{-1}r_b(g)h$.

(35.1) LEMMA: *Given a point* $b \in B$ *there is an open neighborhood* V *of* b *such that for any* $b' \in V$ *there is an* $h \in H$ *such that*

$$r_{b'}(g) = h^{-1}r_b(g)h$$

for all $g \in L$.

PROOF: A right action of $L \times H$ on B is given by $b(g,h) = g^{-1}bh$. We denote by $\Delta_b \subset L \times H$ the isotropy subgroup at $b \in B$; that is, the subgroup of all pairs (g,h) with $b(g,h) = g^{-1}bh = b$. Obviously $gb = bh$ so $h = r_b(g)$ and Δ_b is just the graph of $r_b : h \to H$. Now according to a theorem in Montgomery-Zippin [MZ] there is an open neighborhood V of b such that if $b' \in V$ then $\Delta_{b'}$ is conjugate in $L \times H$ to a subgroup of Δ_b. Thus there is (g_1,h_1) in $L \times H$ with

$$(g_1^{-1},h_1^{-1})\Delta_{b'}(g_1,h_1) \subset \Delta_b.$$

Thus for each $g \in L$, $(g_1^{-1}gg_1, h_1^{-1}r_b(g)h_1) \in \Delta_b$

or

$$r_b(g_1^{-1}gg_1) = h_1^{-1}r_{b'}(g)h_1.$$

Therefore, setting $h = r_b(g_1)h_1^{-1}$ we have $r_{b'}(g) = h^{-1}r_b(g)h$ for all $g \in L$ as required. ∎

If we replace V by $VH = \{b'h | h \in H\}$ then for any pair b,b' in VH we find that the corresponding homomorphisms are conjugate in H. If we cover X by the open sets $\rho(V)$ then by a chaining argument we find

(35.2) LEMMA: *If* $X = B/H$ *is connected then for any pair* b,b' *in* B *there is an* $h \in H$ *such that*

$$r_{b'}(g) = h^{-1} r_b(g) h$$

for all $g \in L$. ∎

From now on we shall assume that for any b,b' in B the homomorphisms $r_b, r_{b'}$ are conjugate in H. As we have just shown this is no restriction if X is connected.

At this point we shall explain how the operator G causes a reduction of structure group H. We fix any point $b_o \in B$ and let $r = r_{b_o}$. The subset $S \subset B$ is defined to be

$$S = \{b \mid b \in B,\ r_b(g) = r(g) \quad \text{for all } g \in L\}.$$

Recalling the action of $L \times H$ on B used in (35.1) the set S consists of those b for which $\Delta_b = \Delta_{b_o}$ and hence S is closed. Furthermore S will meet every H-orbit for given any $b \in B$ there is an $h \in H$ with

$$r(g) = h^{-1} r_b(g) h = r_{bh}(g)$$

so that $bh \in S$.

Let $C(r) \subset H$ be the centralizer of the image of $r : L \to H$. We maintain that $S \cap Sh \neq \emptyset$ if and only if $S = Sh$ and $h \in C(r)$. That is, if $b \in S$ and $bh \in S$ then

$$r(g) = r_{bh}(g) = h^{-1} r_b(g) h = h^{-1} r(g) h$$

for all $g \in L$ and hence $h \in C(r)$.

(35.3) THEOREM: *Suppose that* (G,B,H) *is a principal bundle with operators and that* $L \subset G$ *is the subgroup of elements leaving every point in* B/H *fixed. If* $r : L \to H$ *is a homomorphism that is conjugate in* H *to every* r_b, $b \in B$, *then the group of the principal space* (B,H) *can be reduced to the subgroup* $C(r) \subset G$.

PROOF: Palais has shown that $S \subset B$ produces such a reduction. That is, consider $B/C(r)$ together with the natural map $B/C(r) \to B/H = X$. This is the bundle with fibre $H/C(r)$ which is associated with (B,H). A reduction of structure group is a cross-section of this bundle. The quotient set $S/C(r) \subset B/C(r)$ meets each fibre in exactly one point to yield the required cross-section. ∎

Actually this is not strong enough for our purposes. Using G we have reduced (B,H) to (S,C(r)), but now we want to know when S is also G-invariant so that the triple (G,B,H) is reduced to the triple (G,S,C(r)).

(35.4) LEMMA: *If* G *is acting effectively on* B *then* S *is* G*-invariant if and only if* $L \subset G$ *is a central subgroup.*

PROOF: Recall that $b \in S$ if and only if $gb = br(g)$ for all $g \in L$. Suppose S is G-invariant, then for any $g_1 \in G$

$$gg_1b = g_1br(g) = g_1gb$$

for $b \in S$ and all $g \in L$. Thus $g^{-1}g_1{}^{-1}g_1b = b$ for all $b \in S$. But then $g^{-1}g_1^{-1}gg_1(bh) = bh$ for all $h \in H$ and since S meets every H-orbit it follows that $g^{-1}g_1^{-1}gg_1$ acts trivially on all of B. Finally, because G is effective on B, $g^{-1}g_1^{-1}gg_1 = 1 \in G$ and hence L is central. The converse is elementary. ■ One more point is contained in

(35.5) EXERCISE: *The action of* G *on* B *is effective if and only if the subgroup* L *acts freely.* ■

We next take up the classification problem for right principal H-spaces with G operating. We say that (G,B,H) is isomorphic to (G,B_1,H) if and only if there is a homeomorphism of B onto B_1 which is both H and G equivariant. To proceed further now we shall need additional hypotheses. We assume L is a central subgroup. We fix a homomorphism $r : L \to H$. We consider only triples (G,B,H) in which

i) L is the subgroup of elements mapping every H-orbit into itself

ii) G/L acts freely on X = B/H

iii) for every $b \in B$ the homomorphism $r_b : L \to H$ is conjugate in H to $r : L \to H$.

The collection of isomorphism classes of all triples satisfying these three conditions with G,L,r and H fixed is denoted by Oper (G,r,H). We denote by $C(r) \subset H$ again the centralizer of the image of r in H. The graph $\Delta \subset G \times C(r)$ of r is then a central subgroup.

(35.6) LEMMA: *The element of* Oper (G,r,H) *are in* 1-1 *correspondence with the isomorphism classes of right principal* $G \times C(r)/\Delta$ *spaces.*

PROOF: Begin with a triple (G,B,H). There is then the set S consisting of all b with $r_b = r$. A right action (S, $G \times C(r)$) is then given by $b(g,h) = g^{-1}(bh) = (g^{-1}b)h$. Note that if $b(g,h) = b$ then $gb = bh$ and since G/L is free on X it would follow that $g \in L$ and $h = r(g)$. That is, $(g,h) \in \Delta$ so that every point in S has the same isotropy group; namely, Δ. There results a principal action $(S, G \times C(r)/\Delta)$. It is easy to see that if (G,B,H) is isomorphic to (G,B_1,H) then $(S, G \times C(r)/\Delta)$ is isomorphic to $(S_1, G \times C(r)/\Delta)$.

Then we must go the other way. So we are given a right principal action $(S, G \times C(r)/\Delta)$ which we find first convert into a triple (G,S,C(r)). The left action of G is defined by composing the antihomomorphism $g \to (g^{-1},e)$ with the quotient homomorphism $G \times C(r) \to G \times C(r)/\Delta$. The right action of C(r) arises similarly from $h \to (e,h)$. The action of C(r) is principal and clearly $(gb)h = g(bh)$.

On $S \times H$ we let C(r) act by $(b,h_1) = (bh,h^{-1}h_1)$ and we set B equal to the quotient $(S \times H)/C(r)$. Denoting by $((b,h_1)) \in B$ the point corresponding to (b,h_1) we can define the triple (G,B,H) by

$$g((b,h_1)) = ((gb,h_1))$$

$$((b,h_1))h = ((b,h_1h)).$$

Certainly the actions of G and H commute. Suppose $g((b,h_1)) = ((b,h_1))h$, then there is a $k \in C(r)$ for which $g^{-1}b = bk$ and $k^{-1}h_1h = h_1$. Thus $g \in L$, $k = r(g)$ and $h = h_1^{-1}r(g)h_1$; that is, $r_{((b,h_1))} = h_1^{-1}rh_1$. Incidentally this shows that H acts freely on B and that G/L acts freely on B/H. It is clear from these remarks that the isomorphism classes of (G,B,H) and $(S, G \times C(r)/\Delta)$ determine each other uniquely. ∎

(35.7) **LEMMA:** *If* $r : L \to H$ *admits an extension to a homomorphism* $R : G \to H$ *whose image still lies in the center of* C(r) *then* $(G \times C(r))/\Delta$ *is isomorphic to* $(G/L) \times C(r)$.

PROOF: Denote by $\nu : G \to G/L$ the quotient homomorphism and then define

$$\mu : G \times C(r) \to (G/L) \times C(r)$$

by

$$\mu(g,h) = (\nu(g), R(g^{-1})h).$$

This is a homomorphism because the image of R lies in the center of C(r). If (g,h) is in the kernel of μ then $g \in L$ and $R(g^{-1})h = r(g^{-1})h = e$ so that $r(g) = h$. Thus Δ is the kernel of μ. In addition μ is an epimorphism since $\mu(g,R(g)h) = (\nu(g),h)$. Hence μ induces the required isomorphism. ∎

Associated with a triple (G,B,H) there is the *intermediate quotient* $X = B/H$ on which G/L acts freely and there is the *final quotient* $Y = X/(G/L)$. What we showed in (35.6) was that to each triple (G,B,H) there is canonically associated a right principal $(G \times C(r))/\Delta$ space over the final quotient Y. Now for a space Y we could look at the homotopy classes of maps of Y into $B((G \times C(r))/\Delta)$. We could then say $Map(Y,B((G \times C(r))/\Delta)$ is in 1-1 correspondence with the equivalence classes of triples (G,B,H) with final quotient Y. More technically we are considering $(G,B,H;k)$ where k is a homeomorphism of the final quotient of (G,B,H) with Y. There is an obvious equivalence relation which will yield $Map(Y,B((G \times C(r))/\Delta)$. Later on we shall use and interpret $MSO_*(B((G \times C(r))/\Delta)$.

Fundamentally our concern is with $H = O(k)$, an orthogonal group. Associated to the triple $(G,B,O(k))$ there is a k-plane bundle $\xi \to X$ with operators. This vector bundle is obtained from $B \times R^k$ by introducing $(b,v) \sim (bh^{-1},hv)$. The resulting equivalence class will be denoted by $((b,v)) \in B \times R^k/O(k) = \xi$. Then G acts on ξ by $g((b,v)) = ((gb,v))$. In particular if $g \in L$ then $g((b,v)) = ((gb,v)) = ((br_b(g),v)) = ((b,r_b(g)v))$. Thus L acts on each fibre in a fashion prescribed by the orthogonal representation $r : L \to O(k)$. An important comment is contained in the following

(35.8) LEMMA: *If the centralizer* $C(r) \subset O(k)$ *is connected then the k-plane bundle* $\xi \to X$ *may be oriented.*

PROOF: If $C(r)$ is connected then $C(r) \subset SO(k)$ and hence the section $S/C(r)$ of $B/C(r) \to X$ will surely induce a section of $B/SO(k) \to X$ and hence an orientation of $\xi \to X$. Indeed this arises canonically once r has been chosen. ∎

As an example consider $L \subset G$ to be a cyclic group of odd order $C_{2n+1} = L$. Let us suppose (L,R^q) is an orthogonal representation for which no vector, other than o, is left fixed by the generator, T, of C_{2m+1}. Then by representation theory we know $q = 2s$. Furthermore, if we put $\lambda = \exp(2\pi i/2m+1)$ then we can uniquely choose non-negative integers $n_1,\ldots,n_m$ so that $n_1 + \ldots + n_m = s$ and such that if (C_{2m+1},C^{n_i}) is defined as T acting as scalar multiplication by λ^i then $\sum_1^m(C_{2m+1}, C^{n_i})$ is orthogonally equivalent to (C_{2m+1}, R^{2s}). Thus canonically we would choose $r : C_{2m+1} \to U(s) \subset O(2s)$ to send the generator T into the diagonal unitary matrix in which λ^i appears with multiplicity n_i. The centralizer $C(r)$ is seen to be

$$C(r) = U(n_1) \times \ldots \times U(n_m).$$

So in this situation the orthogonal 2s-plane bundle $\xi \to X$ would be split up canonically into a sum $\xi_1 \oplus \ldots \oplus \xi_m \to X$ of unitary bundles. The dimension of the fibre of ξ_i is n_i and L acts on the fibres of ξ_i as multiplication by λ^i. Notice here that we permit a o-dimensional bundle to occur.

We may inquire as to how the discussion fits into transformation groups. Suppose (G, W^{p+q}) is a smooth action of a finite group on a compact manifold. Let us also suppose $L \subset G$ is a central subgroup which occurs as a *maximal isotropy group* in this action. If $F(L) \subset W$ is the set of fixed points under L then, because L is normal, F(L) is G-invariant. Surely L acts trivially on F(L) and we claim by the maximality of L as an isotropy subgroup that the induced action (G/L,F(L)) is principal (free). Take now a p-dimensional submanifold $X^p \subset F(L)$ which is a union of p-dimensional components of F(L), G-invariant and for which X^p/G is connected. At every point in X^p we receive an orthogonal representation of L in the normal fibre. These are all conjugate in O(q). No non-trivial vector is fixed under all of L. If (G,W) is effective and W/G connected then the representation of L is faithful.

We would then seek some canonical choice of $r : L \to O(q)$ to effect the reduction of structure group of the normal bundle to X^p. We saw how this can be made for L cyclic of odd order. More generally it is possible to work with L a finite abelian group of odd order.

We shall frequently refer to the discussion of this section. We suggest the reader review it now and keep it in mind for later uses.

36. The structure of $MSO_*(C_p)$

This, together with the following two sections, will derive the additive structure of $MSO_*(C_p)$, p an odd prime. A more detailed analysis of the MSO_*-module structure will be postponed for some time. Although these three sections might just as well be combined into one, we have maintained the separate structure from the first edition so that we may give particular emphasis to the fixed point formulas we need to study the free case.

We denote by (T, V^n) an orientation preserving fixed point free diffeomorphism of odd prime period p on a closed oriented manifold. Then (T, V^n) defines an element $[T, V^n] \in MSO_n(C_p)$. Our first important example is the action (T, S^{2k+1}) defined as follows. We put

$$S^{2k+1} = \left\{(z_1,\dots,z_{k+1}) \mid \sum z_i\bar{z}_i = 1\right\}$$

and $\lambda = \exp(2\pi i/p)$ so that

$$T(z_1,\dots,z_{k+1}) = (\lambda z_1,\dots,\lambda z_{k+1}).$$

The quotient lens space S^{2k+1}/T will be denoted simply by $L(2k+1)$. We may form the union $E = \cup S^{2k+1}$ which, given the CW-topology, becomes a principal, contractible, C_p-complex. The quotient $E/T = B(C_p)$ is correspondingly the union of the lens spaces $\cup L(2k+1)$.

As the reader knows, $H_o(B(C_p);Z) \simeq Z$, $H_{2j}(B(C_p);Z) = o$, $j > o$ and $H_{2j+1}(B(C_p);Z) \simeq Z/pZ$. Furthermore the generator of $H_{2k+1}(B(C_p);Z)$ may be taken to be the image of the fundamental class of $L(2k+1) \subset B(C_p)$ under the homomorphism induced by inclusion. Thus it immediately follows that the Thom homomorphism $\mu : MSO_*(C_p) \to H_*(B(C_p);Z)$ is an epimorphism, and hence from (15.1) we have

(36.1) LEMMA: *The oriented bordism spectral sequence for* $B(C_p)$ *collapses.* ∎

There is also a reduced bordism spectral sequence associated with $\widetilde{MSO}_*(B(C_p)) \simeq \widetilde{MSO}_*(B(C_p),pt)$, and of course it too collapses. Note by definition that $[T,V^n] \in \widetilde{MSO}_n(C_p)$ if and only if $[V^n/T] = o$ in MSO_n. From (20.6) we know, however, that $p[V^n/T] = [V^n]$ and since MSO_* has no odd torsion it follows that $[T,V^n] \in \widetilde{MSO}_n(C_p)$ if and only if $[V^n] = o$.

(36.2) LEMMA: *The group* $\widetilde{MSO}_n(C_p)$ *is trivial for* n *even, while for* n *odd it has order* p^t *where* $t = \sum_{j \le n}$ *rank* MSO_j.

PROOF: This is a consequence of the collapsing of the reduced bordism spectral sequence. There is a filtration $o \subset J_{o,n} \subset J_{1,n-1} \subset \dots \subset J_{n,o} = \widetilde{MSO}_n(C_p)$ for which $J_{r,s}/J_{r-1,s+1} \simeq \tilde{H}_r(C_p;MSO_s) \simeq \tilde{H}_r(C_p;Z) \otimes MSO_s$. Hence $J_{2j,s} = J_{2j-1,s+1}$ and the order of $J_{2j+1,s}/J_{2j-1,s+2}$ is $p^{\operatorname{rank} MSO_s}$. The assertion follows. The rank of MSO_s is o if $s \neq o \pmod 4$. If $s = 4r$ then rank MSO_{4r} is equal to the number of partitions of r. ∎

According to our previous remarks there is a choice of $\left\{[T,X^{2k+1}]\right\}_o^\infty$ in $\widetilde{MSO}_*(C_p)$ so that the $\left\{\mu[T,X^{2k+1}]\right\}$ generate $\tilde{H}_*(C_p;Z)$. For example we could use the (T,S^{2k+1}) for this purpose. The next remark is clear from the proof of (19.1).

(36.3) LEMMA: *Suppose* $[T,X^{2k+1}]$, $o \le k < \infty$, *is a sequence of elements in* $\widetilde{M}SO_*(C_p)$ *such that for each* k, $\mu[T,X^{2k+1}]$ *is a generator of* $H_{2k+1}(C_p;Z)$. *Then* $\left\{[T,X^{2k+1}]\right\}$ *is a homogeneous generating set of* $\widetilde{M}SO_*(C_p)$ *as an* MSO_**-module.* ∎

Indeed it may be seen that $J_{2k+1,*}$ is the submodule generated by $\left\{[T,X^{2k+1}]\right\}_o^k$.

We might suggest an exercise at this point. The Thom homomorphism induces an isomorphism

$$\mu : \widetilde{M}SO_1(C_p) \simeq H_1(C_p;Z).$$

Using this show

(36.4) EXERCISE: *If, for* $o < j < p$, $[T_j,S^1]$ *is defined by* $T_j(z) = \lambda_z^j$ *then*

$$j[T_j,S^1] = [T_1,S^1] = [T,S^1]. \quad ∎$$

We must introduce the cohomology ring $H_*(B(C_p);Z/p) \simeq H^*(C_p;Z/pZ)$. We choose $d_1 \in H^1(C_p;Z/pZ)$ so that if $i : S^1/T \to B(C_p)$ is inclusion then $\langle d_1, i_*(\sigma)\rangle = 1 \in Z/pZ$. We then take $d_2 \in H^2(C_p;Z/pZ)$ to be the mod p reduction of the image of d_1 under the integral Bockstein homomorphism $\delta : H^1(C_p;Z/pZ) \to H^2(C_p;Z)$; that is, d_2 is the mod p Bockstein of d_1. Now we can say the well known

(36.5) LEMMA: *For every* r, $H^r(C_p;Z/pZ) \simeq Z/pZ$. *If* $r = 2j$ *then* d_2^j *generates* $H^{2j}(C_p;Z/pZ)$ *while* $d_1d_2^j$ *generates* $H^{2j+1}(C_p;Z/pZ)$. ∎

We shall just write $d_{2j} = d_2^j$ and $d_{2j+1} = d_1d_2^j$. Now the reader will recognize that as long as p is an odd prime we can speak of the mod p Pontrjagin classes of an orthogonal vector bundle and these of course are the mod p reductions of the integral Pontrjagin cohomology classes. Thus a closed oriented manifold has mod p Pontrjagin numbers which are simply the integral Pontrjagin numbers read modulo p. We shall use the following

(36.6) LEMMA: *Suppose* M^n *is a closed oriented manifold with* $n < 2p-2$. *Then* $[M^n]$ *is divisible by* p *in* MSO_n *if and only if every mod* p *Pontrjagin number of* M^n *vanishes.* ∎

Actually we shall comment that Milnor $[M_5]$ showed, for an odd prime, $MSO_*/pMSO_*$ is a polynomial ring over Z/pZ with a generator in each dimension 4j. Further, for $1 \le j \le p-3/2$ the generator corresponds to a manifold X^{4j} for which the Pontrjagin number $s_j(X^{4j}) \neq o \pmod p$. From this (36.6) will follow. There will be more about this later. The lemma is definitely false if $n \geq 2p-2$.

Using $H^*(X;Z/pZ)$ we can associate Pontrjagin numbers modulo p to a singular manifold $f : M^n \to X$ and these again will be bordism invariants of $[M^n,f]$. Take $MSO_*(C_p)$ as the example at hand. Then $\mu[T,X^{2k+1}] \neq o \in H_{2k+1}(C_p;Z)$ if and only if the mod p Pontrjagin number associated to $d_{2k+1} \in H^{2k+1}(C_p;Z/pZ)$ is non-zero. More generally given (T,V^n) we denote by $p_i \in H^{4i}(V^n/T;Z/pZ)$ the mod p Pontrjagin classes of the quotient and by $d_j \in H^j(V^n/T;Z/pZ)$ the image of $d_j \in H^j(C_p;Z/pZ)$ under the homomorphism induced by the homotopically unique map of V^n/T into $B(C_p)$. Then $\langle p_{i_1} \dots p_{i_r} d_j, \sigma_n\rangle$ is a mod p characteristic number of (T,V^n) where $4i_1 + \dots + 4i_r + j = n$.

(36.7) LEMMA: *If* $n < 2p-2$ *and* $[T,V^n]$ *lies in* $\widetilde{MSO}_n(C_p)$ *then* $[T,V^n] = o$ *if and only if all its mod* p *characteristic numbers vanish.*

PROOF: We fix a generating set $\left\{[T,X^{2k+1}]\right\}$ for $\widetilde{MSO}_*(C_p)$. There is, for n odd, an epimorphism $\sum_{4i\leq n} MSO_{4i} \to \widetilde{MSO}_n(C_p)$ which takes $[M^{4i}]$ to $[T,X^{n-4i}][M^{4i}]$. If we copy the proof of (17.2) using these numbers $\langle p_{i_1} \dots p_{i_r} d_j, \sigma\rangle$ mod p then if $([M^o],[M^4],\dots,[M^{4i}],\dots)$ is in the kernel of this epimorphism then for each i, $o \leq i \leq [n/4]$, all the mod p Pontrjagin of $[M^{4i}]$ will vanish. Appealing to (36.6) it follows that if $n < 2p-2$ the kernel of

$$\textstyle\sum_{4i\leq n} MSO_{4i} \to \widetilde{MSO}_n(C_p) \to o$$

is entirely contained in $\sum_{4i\leq n} pMSO_{4i}$. In fact the kernel is exactly $\sum_{4i\leq n} pMSO_{4i}$ because the order of $\widetilde{MSO}_n(C_p)$ is p^t with $t = \sum_{4i\leq n} \text{rank } MSO_{4i}$ while the order of $MSO_{4i}/pMSO_{4i}$ is $p^{\text{rank } MSO_{4i}}$. This completes (36.7). ∎

The result is false if $n \geq 2p-1$. For example, $[T,S^1][CP(p-1)]$ has all its mod p characteristic numbers vanishing, but by the next remark it is non-zero in $\widetilde{MSO}_{2p-1}(C_p)$.

(36.8) LEMMA: *Suppose that* (T,X^{2n+1}) *has* $\mu[T,X^{2n+1}] \neq o$ *in* $H_{2n+1}(C_p;Z/pZ)$. *If* $[M^m] \notin p\,MSO_m$ *then* $[T,X^{2n+1}][M^m] \neq o$ *in* $\widetilde{MSO}_{2n+1+m}(C_p)$.

PROOF: Consider again the reduced bordism spectral sequence for $B(C_p)$. The edge homomorphism $J_{2n+1,o} \to E^2_{2n+1,o}$ maps $[T,X^{2n+1}]$ into some non-zero element γ_{2n+1} of $H_{2n+1}(C_p;Z)$. Since $E^2_{2n+1,o} \otimes MSO_m \simeq E^2_{2n+1,m}$ we see that $\gamma_{2n+1} \otimes [M^m] = o$ if and only if $[M^m] \in p\,MSO_m$. Thus $[T,X^{2n+1}][M^m] \neq o$ if $[M^m] \notin p\,MSO_m$. ■

Next, by analogy with Section 24, we shall define Smith homomorphisms $MSO_n(C_p) \to MSO_{n-2}(C_p)$. We take (T_j,S^{2k+1}) to be $T_j(z_1,\ldots,z_{k+1}) = (\lambda^j z_1,\ldots,\lambda^j z_{k+1})$ where $\lambda = \exp(2\pi i/p)$ and $o < j < p$. Regard $S^{2k-1} \subset S^{2k+1}$ as the set $(o,z_2,\ldots,z_{k+1})$.

We need to recall a remark on transverse regularity. Suppose $\varphi : V^n \to M^m$ is a smooth map between closed oriented manifolds and suppose that φ is transverse regular on a closed oriented submanifold $W^{m-k} \subset M^m$. Then $\varphi^{-1}(W^{m-k}) = V_1^{n-k} \subset V^n$ is a closed *oriented* submanifold. First we orient the normal bundle to W^{m-k} so that the orientation of this normal bundle followed by the orientation of the tangent bundle to W^{m-k} agrees with the orientation on the bundle over W^{m-k} induced from the tangent bundle to M^m. Next orient the normal bundle to $V_1^{m-k} \subset V^n$ so that φ preserves orientation on normal bundles. Finally, choose the orientation on the tangent bundle to V_1^{n-k} so that the orientation of the normal bundle followed by the orientation of the tangent bundle agrees with the restriction of the orientation of V^n.

(36.9) LEMMA: *Given* (T,V^n) *and* $2k + 1 > n$ *there is a smooth equivariant map* $\varphi : (T,V^n) \to (T_j,S^{2k+1})$ *which is transverse regular on* S^{2k-1}. *Let* $W^{n-2} = \varphi^{-1}(S^{2k-1})$ *and let* $T' = T|W^{n-2}$ *and orient* W^{n-2} *as above. The correspondence* $\Delta_j[T,V^n] = [T',W^{n-2}]$ *is a well defined homomorphism of* $MSO_n(C_p)$ *into* $MSO_{n-2}(C_p)$.

PROOF: The proof is entirely like the discussion of the Smith operator in Section 24. Furthermore, Δ_j is an MSO_*-module homomorphism of degree -2. ■

(36.10). LEMMA: *For any odd prime,* $\widetilde{MSO}_*(C_p)$ *is the submodule of* p-*torsion in* $MSO_*(C_p)$.

PROOF: From the collapsing of the reduced bordism spectral sequence for $B(C_p)$ it follows that $\widetilde{MSO}_*(C_p)$ is entirely p-torsion. Now (36.10) will follow because $MSO_*(C_p) \simeq MSO_* \oplus \widetilde{MSO}_*(C_p)$ and MSO_* has no p-torsion. ■
Now since Δ_j is a homomorphism, Δ_j takes $\widetilde{MSO}_n(C_p)$ into $\widetilde{MSO}_{n-2}(C_p)$.

(36.11) THEOREM: *The homomorphism* $\Delta_j : \widetilde{MSO}_n(C_p) \to \widetilde{MSO}_{n-2}(C_p)$ *is an epimorphism. If* n *is even both groups are trivial. If* $n \equiv 3 \pmod 4$ *then* Δ_j *is an isomorphism. If* $n = 4m + 1$ *then*

$$o \to MSO_{4m}/pMSO_{4m} \to MSO_{4m+1}(C_p) \xrightarrow{\Delta_j} \widetilde{M}SO_{4m-1}(C_p) \to o$$

is short exact.

PROOF: If we use $\{[T_j,S^{2k+1}]\}_o^\infty$ as the MSO_*-module generating set then $\Delta_j[T_j,S^{2k+1}] = [T_j,S^{2k-1}]$ and it is immediate that Δ_j is an epimorphism. We compare the orders of $\widetilde{M}SO_{4m+3}(C_p)$ and $MSO_{4m+1}(C_p)$ to see that Δ_j is an isomorphism when $n \equiv 3 \pmod 4$. Comparing orders of $\widetilde{M}SO_{4m+1}(C_p)$ and $\widetilde{M}SO_{4m-1}(C_p)$ shows in this case that the kernel of Δ_j has order $p^{\text{rank } MSO_{4m}}$, which is just the order of $MSO_{4m}/pMSO_{4m}$. Now $[M^{4m}] \to [T_j,S^1][M^{4m}]$ maps MSO_{4m} into the kernel of Δ_j since $\Delta_j[T_j,S^1] = o$. However, $p[T_j,S^1] = o$ and combining this with (36.8) we obtain

$$o \to MSO_{4m}/pMSO_{4m} \to MSO_{4m+1}(C_p).$$

The image is exactly the kernel of Δ_j. ∎

37. The fixed point set of C_p

In this section (T,M^n) will denote an orientation preserving diffeomorphism, possibly with fixed points, for which $T^p =$ identity, p an odd prime. We want to show that the normal bundle to the set of fixed points determines $[M^n]$ modulo $pMSO_n$.

(37.1) LEMMA: *If* (T_j,S^1) *is defined by* $T_j z = \lambda^j z$ *then*

$$[T_j \times T,\ S^1 \times M^n] = [T_j,S^1][M^n]$$

in $\widetilde{M}SO_{n+1}(C_p)$.

PROOF: Actually we shall restrict our attention to (T_1,S^1) given by $T_1 z = \lambda z$. We have already pointed out the universal space $(T,E(C_p))$ in which (T_1,S^1) is the 1-skeleton. The filtration term $J_{1,n}$ in $\widetilde{M}SO_{n+1}(C_p)$ is exactly the image of

$$j_* : \widetilde{M}SO_{n+1}(S^1/T_1) \to \widetilde{M}SO_{n+1}(B(C_p))$$

where $j : S^1/T_1 \to B(C_p)$ is the inclusion. Now S^1/T_1 is again a circle which we denote by $\hat{S}^1$. Then $MSO_n \simeq \widetilde{M}SO_{n+1}(\hat{S}^1)$ by the isomorphism $[V^n] \to [S^1,\text{id}][V^n]$. Under j_* the element $[\hat{S}^1,\text{id}]$ is carried into $[T_1,S^1] \in \widetilde{M}SO_1(C_p)$. Therefore, as in (36.11), the kernel of $\widetilde{M}SO_{n+1}(\hat{S}^1) \to \widetilde{M}SO_{n+1}(C_p)$ is identified with $pMSO_n$.

There is also a geometric interpretation of the isomorphism $\widetilde{M}SO_{n+1}(\hat{S}^1) \simeq MSO_n$. If $f : W^{n+1} \to \hat{S}^1$ is a map of a closed oriented bounding manifold into $\hat{S}^1$ we can assume f is smooth. Choose any regular value $z_0 \in \hat{S}^1$ and send $[W^{n+1},f]$ into $[f^{-1}(z_0)] \in MSO_n$. Here $f^{-1}(z_0)$ is a closed oriented n-manifold in W^{n+1}. Transverse regularity arguments show this is well defined. It is clearly an epimorphism and (17.6) can be used to show trivial kernel.

Consider now the diagonal action $(T_1 \times T, S^1 \times M^n)$. The map

$$f : (S^1 \times M^n)/(T_1 \times T) \to \hat{S}^1$$

is a fibre map with fibre M^n and structure group C_p. Furthermore, $p[(S^1 \times M^n)/(T_1 \times T)] = [S^1 \times M^n] = 0$ and every value of f is regular with $f^{-1}(z_0) = M^n$. But there is also the projection map $\hat{S}^1 \times M^n \to \hat{S}^1$ for which $p^{-1}(z_0)$ is also M^n and thus

$$[(S^1 \times M^n)/(T_1 \times T),f] = [\hat{S}^1,id][M^n]$$

in $\widetilde{M}SO_{n+1}(\hat{S}^1)$. Applying j_* completes the proof. ∎

We continue to consider (T,M^n). We suppose T is an isometry with respect to a suitable Riemannian metric. There is then a normal bundle $\eta \to F$ over the set of fixed points in (T,M^n). The disk bundle $D(\eta) \to F$ can be identified with a closed T-invariant normal tube $N \supset F$. We may think of $T|\dot{N}$ as $(T,S(\eta))$, a fixed point free map of period p on a closed oriented (n-1)-manifold. Furthermore, just as in (22.1), $[T,S(\eta)] = 0 \in \widetilde{M}SO_{n-1}(C_p)$.

Next consider the Whitney sum $C \oplus \eta \to F$ with a trivial complex line bundle, then define $T_1 \times T$ on $C \oplus \eta$ by $(z,v) \to (\lambda z,Tv)$. Thus $(T_1 \times T, S(C \oplus \eta))$ is also fixed point free.

(37.2) THEOREM: *For any* (T,M^n) *we have*

$$[T_1 \times T, S(C \oplus \eta)] = [T_1,S^1][M^n]$$

in $\widetilde{M}SO_{n+1}(C_p)$.

PROOF: The reader may wonder about summing a complex line bundle with a real line bundle. There are two possible approaches here. First replace the trivial complex line bundle with a trivial real 2-plane bundle and use $\begin{pmatrix} \cos(2\pi/p) & 0 \\ 0 & \sin(2\pi/p) \end{pmatrix}$ to define (T_1,R^2). This will do for our purpose here. On the other hand the discussion in Section 35 can be used to impart a complex structure to the normal bundle of every component of the fixed point set and thus $\eta \to F$ becomes a complex vector bundle. We shall take this up in more detail later.

We let $D^2 = \{z \mid |z| \leq 1\}$ be the unit 2-disk and define (T_1, D^2) by $T_1(z) = \lambda z$. Consider then $(T_1 \times T, D^2 \times M^n)$ and $(T_1 \times \mathrm{id}, D^2 \times M^n)$. In the first case the fixed set is identified with $F \subset M^n$ and its new normal bundle is $C \oplus \eta \to F$. In the second case, M^n is the fixed set with normal bundle the trivial complex line bundle $C \to M^n$. Now observe

$$(T_1 \times T, (D^2 \times M^n)^{\cdot}) = (T_1 \times T, S^1 \times M^n)$$

$$(T_1 \times \mathrm{id}, (D^2 \times M^n)^{\cdot}) = (T_1 \times \mathrm{id}, S^1 \times M^n).$$

In view of (37.1) there is a fixed point free (τ, B^{n+2}) for which $(\tau, \dot{B}^{n+2})$ is the disjoint union

$$(T_1 \times T, S^1 \times M^n) \sqcup (T_1 \times \mathrm{id}, -S^1 \times M^n).$$

By taking

$$D^2 \times M^n \cup -B^{n+2} \cup -D^2 \times M^n = V^{n+2}$$

we obtain $a(\tau, V^{n+2})$ on a closed oriented manifold with fixed set $F \sqcup -M^n$. Now (37.2) will follow because

$$[T_1 \times T, S(C \oplus \eta)] - [T_1, S^1][M^n] = 0$$

in $\widetilde{MSO}_{n+1}(C_p)$. ∎

We might note the implications of the theorem. According to (36.8) the right side determines $[M^n]$ in $MSO_n/pMSO_n$. On the other hand, the left side comes directly from the fixed set, its normal bundle and the action of T on that normal bundle. This theorem is an analogue of (22.2).

38. Additive structure of $\widetilde{MSO}_*(C_p)$

Our initial objective in this section is a proof of

(38.1) THEOREM: *If* $[T, X^{2n-1}]$ *in* $\widetilde{MSO}_{2n-1}(C_p)$ *is an element for which* $\mu[T, X^{2n-1}] \neq 0$ *in* $H_{2n-1}(C_p; Z)$ *then* $[T, X^{2n-1}]$ *has order* p^{a+1} *where* $a \cdot (2p-2) < 2n-1 < (a+1) \cdot (2p-2)$.

We must make a preliminary construction. On $CP(p-1)$, complex projective space, we define a map of period p by

$$T[z_1, \ldots, z_p] = [z_1, \lambda z_2, \ldots, \lambda^{p-1} z_p].$$

This has exactly p fixed points; namely, $x_i = [0,\dots,0,1_i,\dots,0]$. To use (37.2) we must put a local coordinate system about each of the x_i. We use

$$(z_1,\dots,z_{p-1}) \to [z_{p-i+1},\dots,z_{p-1},1,z_1,\dots,z_{p-i}]$$

in a neighborhood of x_i. This will be equivariant with respect to the periodic map on C^{p-1} defined by $(z_1,\dots,z_{p-1}) \to (\lambda z_1,\dots,\lambda^{p-1}z_{p-1})$. We denote by $S^{2p-3} \subset C^{p-1}$ the unit sphere. Then $T|S^{2p-3}$ is $(z_1,\dots,z_{p-1}) \to (\lambda z_1,\dots,\lambda^{p-1}z_{p-1})$ at each fixed point x_i in $(T,CP(p-1))$. To use (37.2) we define (T',S^{2p-1}) by $T'(z_1,\dots,z_p) = (\lambda z_1,\lambda z_2,\dots,\lambda^{p-1}z_p)$ Then we have

$$p[T',S^{2p-1}] = [T_1,S^1][CP(p-1)].$$

Since $[CP(p-1)]$ is not divisible by p it follows that $p[T',S^{2p-1}] \neq 0$ from (38.6). However, $p^2[T',S^{2p-1}] = p[T_1,S^1][CP(p-1)] = 0$ so the order of $[T',S^{2p-1}]$ in $\widetilde{M}SO_{2p-1}(C_p)$ is p^2. We proceed to the general case.

From (36.7) we know that $[T,X^{2n-1}]$ has order p if $2n-1 < 2p-2$. Let (T,S^{2k-1}) be an orthogonal fixed point free map of period p, and assume $2k-1 < 2p-4$. Since $p[T,S^{2k-1}] = 0$ there is a (τ,V^{2k}), an orientation preserving diffeomorphism of period p on a closed oriented 2k-manifold, with exactly p fixed points. Around each fixed point there is an invariant S^{2k-1} with $\tau|S^{2k-1} = (T,S^{2k-1})$. In addition $[V^{2k}] \in pMSO_{2k}$ for otherwise we could use (37.2) to find some $[T',S^{2k+1}]$ with order larger than p although $2k+1 < 2p-2$. For each $2k-1 < 2p-4$ select a (T,S^{2k-1}) and a corresponding (τ,V^{2k}).

Consider now a dimension $2n-1 = a(2p-2)-1$. This time take (τ,V^{2m}) to be the a-fold cartesian product of $CP(p-1)$ with itself together with the diagonal action arising from our original $(T,CP(p-1))$. Now (τ,V^{2m}) has exactly p^a fixed points each surrounded by an invariant sphere for which $\tau|S^{2n-1}$ is independent of the fixed point. If (T,S^{2n-1}) is this common orthogonal map of period p on the spheres about the fixed points, then by (37.2) we can find an orthogonal fixed point free (T',S^{2n+1}) with

$$p^a[T',S^{2n+1}] = [T_1,S^1][CP(p-1)]^a.$$

We recall $[M_5]$ that $MSO_*/pMSO_*$ is a polynomial algebra over Z/pZ with a generator in each dimension divisible by 4 and $CP(p-1)$ may be used as the $2p-2$ dimensional generator. Therefore $[CP(p-1)]^a$ is not divisible by p and hence $[T',S^{2n-1}]$ has order p^{a+1}. So far we have shown (38.1) in dimensions less that $2p-2$ and for a certain generator in each dimension of the form $a(2p-2)+1$.

Consider next $2n+1 = a(2p-2)+k$, $1 \leq k < 2p-4$ and k odd. There is then $V^{2n+2} = (CP(p-1))^a \times V^{k+1}$. There are already maps of period p on both factors so let (τ', V^{2n+2}) be the resulting diagonal map. Then τ' has p^{a+1} fixed points and all have equivariantly diffeomorphic neighborhoods. Let (T,S^{2n+1}) denote the common fixed point free orthogonal action surrounding each fixed point. Applying (37.2) to (τ', V^{2n+2}) we obtain a (T',S^{2n+3}) with

$$o = p^{a+1}[T',S^{2n+3}] = [T_1,S^1][CP(p-1)]^a[V^{k+1}].$$

The above is o because $[V^{k+1}]$ is divisible by p. Hence the order of $[T',S^{2n+3}]$ does divide p^{a+1} where $a(2p-2)+3 \leq 2n+3 < (a+1)(2p-2)$.

At this point we have for each n a particular orthogonal free action of period p such that

i) $[T',S^{2n-1}]$ has order p^{a+1} for $2n-1 = a(2p-2)+1$

ii) $[T',S^{2n-1}]$ has order dividing p^{a+1} for

$a(2p-2)+3 \leq 2n-1 < (a+1)(2p-2)$.

These particular $[T',S^{2n-1}]$ are a generating set for $\widetilde{M}SO_*(C_p)$.

Let $[T,X^{2n-1}]$, $2n-1 = a(2p-2)+1$, be an element for which $\mu[T,X^{2n-1}] \neq o \in H_{2n-1}(C_p;Z)$. Using (36.3) there is an integer b, $o < b < p$, such that $[T,X^{2n-1}] = b[T',S^{2n-1}] + [T',S^{2n-5}][V^4] + \ldots$ Multiplying through by p^a we find that

$$p^a[T,X^{2n-1}] = bp^a[T',S^{2n-1}] = b[T_1,S^1][CP(p-1)]^a$$

and hence $[T,X^{2n-1}]$ has order p^{a+1}.

We have to show that if $a(2p-2)+3 \leq 2n-1 < (a+1)(2p-2)$ then $[T,X^{2n-1}]$ still has order p^{a+1}. Again we write $[T,X^{2n-1}] = b[T',S^{2n-1}] + \ldots$ to see that the order of $[T,X^{2n-1}]$ divides p^{a+1}. Now with $d_2 \in H^2(C_p;Z) \simeq Z/pZ$ the diagram

$$\begin{array}{ccc} \widetilde{M}SO_{2n-1}(C_p) & \xrightarrow{\mu} & H_{2n-1}(C_p;Z) \\ \downarrow \Delta & & \downarrow \cap d_2 \\ \widetilde{M}SO_{2n-3}(C_p) & \xrightarrow{\mu} & H_{2n-3}(C_p;Z) \end{array}$$

commutes. Thus if Δ is applied to $[T,X^{2n-1}]$ exactly $(n-a(p-1)-1)$ times we obtain a generator in $\widetilde{M}SO_{a(2p-2)+1}(C_p)$ which we then know has order p^{a+1}. The proof of (38.1) is completed. ■ We explicitly point out again

(38.2) LEMMA: *With notation as in* (38.1), *when* $2n-1 = a(2p-2)+1$ *then*

$$p^a[T,X^{2n-1}] = b[T_1,S^1][CP(p-1)]^a$$

where $o < b < p$. ■

Recall again that MSO_*/Tor is a polynomial ring over Z with generators $[Y^{4k}]$, $1 \leq k < \infty$, in MSO_{4k}. For an odd prime we select $Y^{2p-2} = CP(p-1)$. We fix an MSO_*-generating set $[T,X^{2n-1}]$ for $\widetilde{M}SO_*(C_p)$ with $\mu[T,X^{2n-1}] \neq o$ and $\Delta[T,X^{2n+1}] = [T,X^{2n-1}]$. Let $\Gamma(p) \subset MSO_*/\mathrm{Tor}$ be the polynomial subring generated by all $[Y^{4k}]$ with $4k \neq 2p-2$.

(38.3) LEMMA: *Suppose*

$$\sum_{i+j=n} [T,X^{4j+1}][M^{4i}] = o$$

with each $[M^{4i}] \in \Gamma_{4i}(p)$. *Then* $[M^{4i}]$ *lies in* $p^{a+1}MSO_{4i}$ *where* $a(2p-2) < 4j+1 < (a+1)(2p-2)$.

PROOF: This is by induction on n. Thus we assume the result for all $m < n$. Then given

$$\text{(i)} \qquad \sum_{i+j=n} [T,X^{4j+1}][M^{4i}] = o$$

we apply Δ^2 to this equation and obtain

$$\text{(ii)} \qquad \sum_{i+j-1=n-1} [T,X^{4(j-1)+1}][M^{4i}] = o.$$

By the inductive hypothesis $[M^{4i}] \in p^{a+1}MSO_{4i}$ if $a(2p-2)+4 < 4_j + 1 < (a+1)(2p-2)$ while $[M^{4i}] \in p^a MSO_{4i}$ if $4_j + 1 = a(2p-2)+3$. The order of $[T,X^{4j+1}]$ is p^{a+1} where $a(2p-2) < 4j+1 < (a+1)(2p-2)$. Thus equation (i) reads

$$\sum_a [T,X^{a(2p-2)+3}][M^{4n-a(2p-2)-2}] = o.$$

We have $[M^{4n-a(2p-2)-2}] = p^a[V^{4n-a(2p-2)-2}]$ or

$$\sum_a p^a[T,X^{a(2p-2)+3}][V^{4n-a(2p-2)-2}] = o.$$

By (38.2) we can write, with $b_a \neq o \pmod p$

$$[T_1,S^1](\sum_a b_a([CP(p-1)])^a[V^{4n-a(2p-2)-2}]) = o.$$

This implies $\sum_a b_a([CP(p-1)])^a[V^{4n-a(2p-2)-2}]$ is divisible by p. Now surely each $[V^{4n-a(2p-2)-2}]$ lies in $\Gamma(p)$ also and hence $[V^{4n-a(2p-2)-2}]$ is contained in $pMSO_*$. Thus $[M^{4n-a(2p-2)-2}]$ is in $p^{a+1}MSO_*$ as required. ∎

We fix an integer $4n+1$. We define a sequence of integers a_i by the rule $a_i(2p-2) < 4(n-i)+1 < (a_i+1)(2p-2)$. Consider next the sum $\sum_o^n \Gamma_{4i}(p)/p^{a_i+1}\Gamma_{4i}(p)$. There is a homomorphism

$$\sum_o^n \Gamma_{4i}(p)/p^{a_i+1}\Gamma_{4i}(p) \to \widetilde{M}SO_{4n+1}(C_p)$$

given by $[M^{4i+1}] \to [T,X^{4(n-i)+1}][M^{4i}]$. This is well defined since p^{a_i+1} is the order of $[T,X^{4(n-i)+1}]$. Then (38.3) is exactly the statement that this is a monomorphism. We would also like to know this is an epimorphism. The order of $\widetilde{M}SO_{4n+1}(C_p)$ is p^t where $t = \sum_{i\leq n} s_i$ and s_i = rank MSO_{4i}; that is, s_i is the number of partitions of i. Let t_i be the number of partitions of i which do not contain $(p-1)/2$. Then $s_i = \sum_a t_{i-a(p-1)/2}$. Hence $\sum_{i\leq n} s_i = \sum_{i\leq n,a} t_{i-a(p-1)/2} = \sum_{j\leq n} c_j \cdot t_j$. We can calculate the c_j as follows. Suppose $4j + b(2p-2) \leq 4n < 4j + (b+1)(2p-2)$, then we get a t_j in the sum for each $i = j + a(p-1/2$, $a = o,1,\ldots,b+1$. Hence $c_j = b+1$ with b as above. A computation of the order of $\sum_o^n \Gamma_{4i}(p)/p^{a_i+1}\Gamma_{4i}(p)$ shows it is equal to $p^{\Sigma c_j t_j}$. Therefore

$$\sum_o^n \Gamma_{4i}(p)/p^{a_i+1}\Gamma_{4i}(p) \simeq \widetilde{M}SO_{4n+1}(C_p).$$

From (36.11) we also recall that

$$\Delta : \widetilde{M}SO_{4n+3}(C_p) \simeq \widetilde{M}SO_{4n+1}(C_p)$$

We can put this together to yield the additive structure of $\widetilde{M}SO_*(C_p)$.

(38.4) THEOREM: *If the* MSO_**-generating set* $[T,X^{2n+1}]$ *is selected then the group* $\widetilde{M}SO_*(C_p)$ *is a direct sum of cyclic groups* $C_{2n+1,k_1,\ldots,k_s}$ *with generators the products* $[T,X^{2n+1}][Y_{4k_1} \ldots Y_{4k_s}]$, *one for each* $n \geq o$ *and each* $(k_1,\ldots,k_s)$ *with all* k_j *different from* $(p-1)/2$. *The order of the generator is* p^{a+1}, $a(2p-2) < 2n+1 < (a+1)(2p-2)$. ∎

39. Cyclic groups of odd order

In the first edition we actually did calculations similar to those of Section 38 for a cyclic group of odd prime power order, however, we extend to the reader the mercy of omitting that discussion and instead we shall content ourselves with some general remarks. If m is odd then there are fixed point free maps of period m (T,S^{2k-1}) given by

$$T(z_1,\dots,z_k) = (\lambda z_1,\dots,\lambda z_k)$$

where $\lambda = \exp(2\pi i/m)$. For each $k \geq 1$ the image of $[T,S^{2k-1}]$ under the Thom homomorphism $\mu : MSO_{2k-1}(C_m) \to H_{2k-1}(C_m;Z) \simeq Z/mZ$ is a generator of this homology group. As a consequence the oriented bordism spectral sequence for $MSO_*(C_m)$ collapses as does the reduced bordism spectral sequence. A sequence of elements $[T,X^{2k-1}]$, $1 \leq k < \infty$, is an MSO_*-module generating set of $\widetilde{M}SO_*(C_m)$ if and only if $\mu[T,X^{2k-1}]$ is a generator of $H_{2k-1}(C_m;Z)$ for all $k \geq 1$.

It is possible to define a Smith operator $\Delta : MSO_*(C_m) \to MSO_*(C_m)$. This will have properties to those discussed in Section 36.

For a general finite group the Thom homomorphism $\mu : MSO_*(G) \to H_*(G;Z)$ may fail to be an epimorphism. In terms of weakly complex bordism, Landweber has established the following definitive result [L].

(39.1) LANDWEBER: *If* G *is a finite group then* $\mu : MU_*(G) \to H_*(G;Z)$ *is an epimorphism if and only if* G *has periodic cohomology.* ∎

40. Families of subgroups

In this section we fix a finite *abelian* group G. We shall consider a family F of subgroups which satisfies the condition that if $K \in F$ and if $K^1 \subset K$ is a subgroup, then $K^1 \in F$ also. All of our families of subgroups will satisfy this condition. We allow the possibility of the empty family.

Consider then a pair of families $F \supset F'$

(40.1) DEFINITION: *An* (F,F')*-free action of* G *is an orientation preserving action of* G *on a compact oriented manifold,* B^n *, such that*

(i) *if* $x \in B^n$ *then the isotropy subgroup at* x, G_x, *belongs to* F

(ii) *if* $x \in \dot{B}^n$ *then* $G_x \in F'$.

We remind the reader that the *isotropy subgroup* is defined by $G_x = \{g \mid gx = x\}$. Notice that if F' is the empty family then necessarily $\dot{B}^n = \emptyset$, so that B^n is closed. In this case we say (G,B^n) is F-free, with F' empty understood.

We wish to define an oriented bordism theory of (F,F')-free actions. The concept of an equivariant orientation preserving diffeomorphism between (F,F')-free actions is clear. Similarly $-(G,B^n) = (G,-B^n)$.

(40.2) DEFINITION: *An* (F,F')*-free action* (G,B^n) *is said to bord if and only if there is an* (F,F)*-free action* (G,W^{n+1}) *for which* (G,B^n) *is equivariantly embedded into* $(G,\dot{W}^{n+1})$ *so that* $B^n \cup B_o^n = \dot{W}^{n+1}$, $\dot{B}^n = B^n \cap B_o^n = \dot{B}_o^n$ *and so that for all* $x \in B_o^n$, $G_x \in F'$.

It is understood that the orientation of B^n is compatible with that of W^{n+1}. We say that (G,B_1^n) is bordant to (G,B_2^n) if and only if their disjoint union $(G,B_1^n) \sqcup (G,-B_2^n)$ bords. Obviously this is a relative version, with restrictions, of $\mathcal{O}_*(G)$. Using the approach of Section 4 this can be shown to define an equivalence relation. Briefly, given (G,B^n) put $W^{n+1} = I \times B^n$ and let G act on W^{n+1} by $g(t,x) = (t,gx)$, which is (F,F)-free. Now $\dot{W}^{n+1} = \dot{I} \times B^n \cup (I \times -\dot{B}^n)$ so that $\dot{I} \times B^n = B^n \sqcup -B^n$ and $B_o^n = I \times -\dot{B}^n$, on which every isotropy group belongs to F'. From this it will follow that (G,B^n) is bordant to itself. For transitivity we would have

$$B^n \sqcup -B_1^n \subset \dot{W}^{n+1}$$
$$B_1^n \sqcup -B_2^n \subset \dot{W}_1^{n+1}$$

then we form $W^{n+1} \cup W_1^{n+1}$ by identifying along B_1^n in both boundaries. This will define a bordism between (G,B^n) and (G,B_2^n).

If $\mathcal{O}_n(G;F,F')$ is the resulting collection of bordism classes then an abelian group structure is imposed as usual by disjoint union. We can also introduce the direct sum $\mathcal{O}_*(G;F,F') = \sum \mathcal{O}_n(G;F,F')$. On this last there is naturally defined an MSO_*-module structure.

Let us point out two special cases. If F is the family of all subgroups and F' is the empty family then very simply we have $\mathcal{O}_*(G)$, the oriented bordism algebra of all orientation preserving actions of G on closed oriented manifolds. At the opposite extreme we could let $F = \{e\}$, the trivial subgroup only. By again choosing F' empty we would then have $MSO_*(G)$. These families were introduced to develop a mechanism for bridging the gap between these two extremes.

Now suppose that we are given any pair of families $F \supset F'$. Any F'-free action is surely also F-free and in fact this inclusion will induce a homomorphism

$$\alpha : \mathcal{O}_n(G;F') \to \mathcal{O}_n(G;F).$$

Similarly any F-free action on a closed manifold is also (F,F')-free and this produces a homomorphism

$$\beta : \mathcal{O}_n(G;F) \to \mathcal{O}_n(G;F,F').$$

Finally a homomorphism

$$\partial : \mathcal{O}_n(G;F,F') \to \mathcal{O}_{n-1}(G;F')$$

is given by $\partial[G,B^n] = [G,\dot{B}^n]$. As the reader should expect, we can combine this into

(40.3) THEOREM: *For any pair of families* $F \supset F'$ *the sequence*

$$\dots \to \mathcal{O}_n(G;F') \xrightarrow{\alpha} \sigma_n(G;F) \xrightarrow{\beta} \mathcal{O}_n(G;F,F') \xrightarrow{\partial} \mathcal{O}_{n-1}(G;F') \to \dots$$

is exact.

We shall first indicate a lemma which is analogous to the remark preceding (5.6).

(40.4) LEMMA: *Let* (G,B^n) *be an* (F,F')*-free action and let* W^n *be a compact manifold regularly embedded in the interior of* B^n *and invariant under* G. *If for every* $x \in B^n \setminus W^n$, $G_x \subset F'$ *then* $[G,W^n] = [G,B^n]$ *in* $\mathcal{O}_n(G;F,F')$.

PROOF: We consider again the action $(G, I \times B^n)$. By straightening the angle, Section 3, equivariantly this product becomes a differentiable manifold on which G acts smoothly. Directly from the definition then it follows that $(G, 1 \times B^n)$ is (F,F')-bordant to $(G, o \times W^n)$. ∎

Taking $W^n = \emptyset$ it follows that the composition

$$\mathcal{O}_n(G;F') \to \mathcal{O}_n(G;F) \to \mathcal{O}_n(G;F,F')$$

is trivial. The rest of the proof is directly analogous to (5.6) and is left as an exercise. ∎

Of course if $F \supset F' \supset F''$ there is an exact sequence

$$\dots \to \mathcal{O}_n(G;F',F'') \to \mathcal{O}_n(G;F,F'') \to \mathcal{O}_n(G;F,F') \xrightarrow{\partial} \mathcal{O}_{n-1}(G;F',F'') \to \dots$$

derived as usual from the exact sequence for pairs of families.

So far our machine lacks a crucial feature. So far we have no relation between an action of G and the actions of the various subgroups of G. In the rest of this section we shall remedy this situation.

Two points must be borne closely in mind. First, any family of subgroups of $K \subset G$ can be thought of also as a family of subgroups of G. Second, if F is a family of subgroups of G and $K \subset G$ then by $K \cap F$ we denote the family of subgroups of K which also belong to F.

If now $K \subset G$ is a subgroup and $F \supset F'$ is a pair of families of subgroups in K we can define an *induction* homomorphism

$$E_{GK} : \mathcal{O}_n(K;F,F') \to \mathcal{O}_n(G;F,F').$$

In general consider an action (K,X) on a space X, then on the product $G \times X$ we define a left principal action of K by $k(g,x) = (gk^{-1},kx)$. We denote the quotient $(G \times X)/K$ simply by $G \times_K X$. An element of this quotient is denoted by $((g,x))$. Now G acts on $G \times_K X$ by $g_1((g,x)) = ((g_1g,x))$. Let us analyse the isotropy subgroup of G at $((g_0,x_0))$. If $g((g_0,x_0)) = ((gg_0,x_0)) = ((g_0,x_0))$ then there is a $k \in K$ with $gg_0 = g_0k^{-1}$ and $kx_0 = x_0$. But then, since G is abelian, $g = k^{-1}$ and $G((g_0,x_0)) = K_{x_0}$.

If X is a compact oriented manifold with boundary on which K acts smoothly as a group of orientation preserving diffeomorphisms then $G \times_K X$ is a disjoint union of a number of copies of X equal to the index of K in G. Furthermore G acts smoothly on $(G \times_K X)$ as a group of orientation preserving diffeomorphisms, and $(G,(G \times_K X)^{\cdot}) = (G, G \times_K \dot{X})$. In fact

$$E_{GK} : \mathcal{O}_n(K;F,F') \to \mathcal{O}_n(G;F,F')$$

given by

$$E_{GK}[K,B^n] = [G,(G \times_K B^n)]$$

is a well defined homomorphism.

This E_{GK} is the bordism analogue of the induced representation in representation theory. Dually, if $F \supset F'$ is a pair of families in G then there is a *restriction homomorphism*

$$R_{KG} : \mathcal{O}_n(G;F,F') \to \mathcal{O}_n(K;\ K \cap F,\ K \cap F').$$

This is simply defined by restricting the action of G to that of the subgroup K.

When (F,F') is a pair of families of subgroups of K then the composition

$$\mathcal{O}_n(K;F,F') \xrightarrow{E} \mathcal{O}_n(G;F,F') \xrightarrow{R} \mathcal{O}_n(K;F,F')$$

is defined. The next result is as to be expected.

(40.5) LEMMA: *If* $\mathcal{F} \supset \mathcal{F}'$ *are a pair of families of subgroups in* $K \subset G$ *then the composition*

$$R_{KG}E_{GK} : \mathcal{O}_n(K;\mathcal{F},\mathcal{F}') \to \mathcal{O}_n(K;\mathcal{F},\mathcal{F}')$$

is multiplication by the index of K *in* G.

PROOF: Consider $G \times_K X$ on which K acts by $k((g,x)) = ((kg,x)) = ((kgk^{-1},kx)) = ((g,kx))$. Let $\nu: G \to G/K$ be the quotient homomorphism, then an equivariant homeomorphism

$$(K, G \times_K X) \to (K, (G/K) \times X)$$

is given as follows. Select $e = g_1, g_2, \ldots, g_r$ as coset representatives of K in G. Then we can write $((g,x)) = ((g_i k,x)) = ((g_i,kx))$ for a unique g_i and a unique $k \in K$. Now

$$((g,x)) \to (\nu(g),kx)$$

is the required homeomorphism. It is equivariant with respect to K only acting on the second factor of $(G/K) \times X$. From these observations (40.5) will immediately follow. ∎

To discuss the other composition it would be appropriate to first mention the natural $\mathcal{O}_*(G)$-module structure on $\mathcal{O}_*(G;\mathcal{F},\mathcal{F}')$. Suppose (G,B^n) is $(\mathcal{F},\mathcal{F}')$-free and (G,M^m) is an orientation preserving action on a closed manifold. We have then the diagonal action $(G, B^n \times M^m)$ given by $g(x_1,x_2) = (gx_1,gx_2)$. At each point of the product the isotropy subgroup is $G_{x_1} \cap G_{x_2}$ so the diagonal action is always $(\mathcal{F},\mathcal{F}')$-free no matter which families $(\mathcal{F},\mathcal{F}')$ are specified. Thus $[G,B^n][G,M^n] = [G, B^n \times M^m]$ yields a well defined graded $\mathcal{O}_*(G)$-module structure on $\mathcal{O}_*(G;\mathcal{F},\mathcal{F}')$.

Suppose now we have a subgroup $K \subset G$. If $X \in \mathcal{O}_*(G;\mathcal{F},\mathcal{F}')$ and $Y \in \mathcal{O}_*(K)$ then the product $R_{KG}(X) \cdot Y \in \mathcal{O}_*(K;\ \mathcal{F} \cap K,\ \mathcal{F}' \cap K)$ is defined.

(40.6) FROEBENIUS RECIPROCITY: In $\mathcal{O}_*(G;\mathcal{F},\mathcal{F}')$

$$E_{GK}(R_{KG}(X) \cdot Y) = X \cdot E_{GK}(Y).$$

PROOF: Represent X by (G,B^n) and Y by (K,M^m). Then express an element in $G \times_K (B^n \times M^m)$ by $((g,x_1,x_2))$ on which G acts by $g_1((g,x_1,x_2)) = ((g_1 g,x_1,x_2))$. Furthermore $((gk^{-1},kx_1,kx_2)) = ((g,x_1,x_2))$. Now we define a G-equivariant diffeomorphism of $G \times_K (B^n \times M^m)$ with $B^n \times (G \times_K M^m)$ by $((g,x_1,x_2)) \to (gx_1,((g,x_2))$. The assertion follows. ∎

Incidentally, if we take $Y = [K,pt]$ this then is the identity element of $\mathcal{O}_*(K)$ and $E_{GK}[K,pt] = [G,G/K]$. Thus for $X \in \mathcal{O}_*(G;F,F')$

$$E_{GK}R_{GK}(X) = X[G,G/K].$$

There are also some naturality rules obeyed by the operators E and R.

(40.7) LEMMA: *If* $F \supset F'$ *are families of subgroups of* $K \subset G$ *then the diagram*

$$\begin{array}{ccccccccc}
\cdots \to & \mathcal{O}_n(K;F') & \to & \mathcal{O}_n(k;F) & \to & \mathcal{O}_n(K;F,F') & \to & \mathcal{O}_{n-1}(K;F') & \to \cdots \\
& \downarrow E_{GK} & & \downarrow E_{GK} & & \downarrow E_{GK} & & \downarrow E_{GK} & \\
\cdots \to & \mathcal{O}_n(G,F') & \to & \mathcal{O}_n(G;F) & \to & \mathcal{O}_n(G;F,F') & \to & \mathcal{O}_{n-1}(G;F') & \to \cdots
\end{array}$$

commutes. On the other hand if $F \supset F'$ *are families of subgroups of* G *then the diagram*

$$\begin{array}{ccccccccc}
\cdots \to & \mathcal{O}_n(G;F') & \longrightarrow & \mathcal{O}_n(G;F) & \longrightarrow & \mathcal{O}_n(G;F,F') & \longrightarrow & \mathcal{O}_{n-1}(G,F') & \longrightarrow \cdots \\
& \downarrow R_{KG} & & \downarrow R_{KG} & & \downarrow R_{KG} & & \downarrow R_{KG} & \\
\cdots \to & \mathcal{O}_n(K;K\cap F') & \to & \mathcal{O}_n(K;k\cap F) & \to & \mathcal{O}_n(K;K\cap F,K\cap F') & \to & \mathcal{O}_{n-1}(K;K\cap F') & \to \cdots
\end{array}$$

also commutes. ∎

The argument is omitted. Next we can show associativity for induction.

(40.8) LEMMA: *Let* G *be a finite abelian group and let* $K \supset L$ *be any pair of subgroups. If* $F \supset F'$ *are families of subgroups of* L *then the diagram*

$$\begin{array}{ccc}
\mathcal{O}_n(L;F,F') & \xrightarrow{E_{KL}} & \mathcal{O}_n(K;F,F') \\
{\scriptstyle E_{GL}} \searrow & & \swarrow {\scriptstyle E_{GK}} \\
& \mathcal{O}_n(G;F,F') &
\end{array}$$

will commute.

PROOF: Consider (L,B^n) and form the product $G \times K \times B^n$. Now $K \times L$ will act on $G \times K \times B^n$ by

$$(k,l)(g,k_1,x) = (gk^{-1},kk_1l^{-1},lx).$$

But then

$$(G \times K \times B^n)/K \times L \simeq (G \times K \times_L B^n)/K \simeq G \times_K B^n).$$

However,

$$G \times K \times B^n/K \simeq G \times B^n$$

under $((q_1, k_1 x)) \to (gk_1, x)$. Moreover this induces

$$((G \times K \times B^n)/K)/L \simeq G \times_L B^n.$$

It is then seen that equivariantly

$$(G, G \times_L B^n) \simeq (G, G \times_K (K \times_K B^n)) \simeq (G, (G \times K \times B^n)/K \times L)$$

and therefore $E_{GL}[L, B^n] = E_{GK}E_{KL}[L, B^n]$. ∎

(40.9) EXERCISE: *State and prove the associativity rule for restriction homomorphisms.* ∎

41. Principal actions

We shall take up consideration of E and R as applied to $MSO_*(G)$ and $MSO_*(K)$. Suppose (G, U) is a left principal contractible G-space so that $U/G = B(G)$. We fix a subgroup $K \subset G$ and consider

$$U \xrightarrow{\nu_1} U/K \xrightarrow{\nu_2} U/G$$

and we set $\nu = \nu_2 \nu_1 : U \to U/G$. Now suppose that $f : X \to B(K)$ is a map, and let $F = \nu_2 f : X \to B(G)$ be the composition of f with ν_2. In the product $U \times X$ we have the subspace

$$Y = \{(u, x) \mid \nu_1(u) = f(x)\}$$

A left principal action of K on Y is given by $(ku, x) = k(u, x)$. There is also

$$Y_1 = \{(u, x) \mid \nu(u) = F(x)\}.$$

Clearly $Y \subset Y_1$ and G acts principally on Y_1. There is a map

$$G \times Y \to Y_1$$

given by $(g, u, x) \to (gu, x)$. This is onto; furthermore for any $k \in K (gk^{-1}, ku, x) \to (gu, x)$. The induced

$$G \times_K Y \to Y_1$$

equivariantly identifies $(G, G \times_K Y)$ with (G, Y_1).

(41.1) LEMMA: *If* $K \subset G$ *is any subgroup then* $E_{GK} : MSO_n(K) \to MSO_n(G)$ *coincides with the homomorphism induced by the map* $B(K) \to B(G)$. *In particular*

$$\begin{array}{ccc} MSO_n(K) & \xrightarrow{E_{GK}} & MSO_n(G) \\ \downarrow \mu & & \downarrow \mu \\ H_n(K;Z) & \xrightarrow{i_{GK}} & H_n(G;Z) \end{array}$$

is a commutative diagram. ∎

The homomorphism $R_{KG} : MSO_n(G) \to MSO_n(K)$ is the analogue of the transfer homomorphism in the homology of finite groups. If $t_{KG} : H_n(G;Z) \to H_n(K;Z)$ denotes this transfer then

(41.2) LEMMA: *The diagram*

$$\begin{array}{ccc} MSO_n(G) & \xrightarrow{R_{KG}} & MSO_n(K) \\ \downarrow \mu & & \downarrow \mu \\ H_n(G;Z) & \xrightarrow{t_{KG}} & H_n(K;Z) \end{array}$$

will commute. ∎

As we shall not use this the proof is omitted. The result with which we are principally concerned is

(41.3) THEOREM: *If* G *is a finite cyclic group of odd order and if* $K \subset G$ *is any subgroup then*

$$E_{GK} : MSO_n(K) \to MSO_n(G)$$

is a monomorphism for n *odd and an isomorphism for* n *even.*

PROOF: First we note that for n odd, $i_{GK} : H_n(K;Z) \to H_n(G;Z)$ is a monomorphism while for n even, $n > 0$, both groups are trivial. We have an isomorphism for $n = 0$. Since E_{GK} is induced by the map $B(K) \to B(G)$ there is induced a natural, compatible homomorphism of the oriented bordism spectral sequences

$${}'E^r \to E^r.$$

Both spectral sequences collapse. At the $r = 2$ level we have

$$i_{GK} \otimes id : H_p(K;Z) \otimes MSO_q \to H_p(G;Z) \otimes MSO_q.$$

This is an isomorphism if $p = o$, a monomorphism if p is odd and if $p > o$ and even both sides vanish.

Then because both spectral sequences collapse the same remark is true of

$$'J_{p,q}/'J_{p-1,q+1} \to J_{p,q}/J_{p-1,q+1}.$$

From this (41.3) will follow. ∎

42. A bordism theory

We go back now to put a bordism interpretation on Section 35. We fix the groups G,H and the central subgroup $L \subset G$ together with a homomorphism $r : L \to H$. In (35.6) the group $(G \times C(r))/\Delta$ was introduced. This has a classifying space, of course, and we are in the oriented bordism MSO_*-module, $MSO_*(B(G \times C(r)/\Delta))$. As it stands this will represent the bordism classification of right principal $(G \times C(r))/\Delta$-spaces over closed oriented manifolds. But, in view of the discussion following the proof of (35.7), another interpretation is possible. Suppose W (to be regarded as the final quotient) is a compact oriented manifold. Then we could consider a triple (G,B,H) over W where, as before, L is the subgroup mapping every H-orbit onto itself and for every $b \in B$ the homomorphism $r_b : L \to H$ is conjugate to r. Furthermore, G/L acts freely on B/H. Notice then that the inclusion $\dot{W} \subset W$ will induce a triple $(G,\dot{B},H)$ over the oriented boundary. It is correct to regard B as a compact manifold. For G finite we think of B/H as oriented so that $B/H \to W$ has degree equal to the order of G/L. Therefore we could directly define the bordism of such triples over closed oriented manifolds. We may denote this by the symbol $MSO_k(G, r : L \to H)$, where k refers to the dimension of the final quotient manifold. The conclusion is that $MSO_k(G, r : L \to H) \simeq MSO_k(B(G \times C(r)/\Delta)$.

Suppose now we alter our viewpoint slightly. Given G,L and H we consider triples (G,B,H) but we put no restrictions on the homomorphisms r_b. We should still obtain a bordism group $MSO_k(G;\ L \to H)$. This will naturally split up into a direct sum as follows. Choose a sequence of homomorphisms $r_j : L \to H$ so that r_j is not conjugate to r_i in H if $i \neq j$, but so that any homomorphism $r : L \to H$ is conjugate in H to one of the r_j. Then

$$MSO_k(G;\ L \to H) \simeq \sum_j MSO_k(G;\ r_j : L \to H).$$

Let us see how this works in practice. Suppose $L = C_m$ is a cyclic group of odd order, and that H is $O(2s)$. We are interested in listing, up to conjugacy, the orthogonal representations of C_m which leave no vector in R^{2s}, other than o, fixed. Put $\lambda = \exp(2\pi i/m)$ and $\mu = (m-1)/2$. To every ordered sequence of non-negative integers $(n_1,\ldots,n_\mu)$ for which $n_1 + \ldots + n_\mu = s$ we associate the diagonal matrix in $U(s)$ in which λ^i appears on the diagonal with multiplicity n_i. If $\tau \in C_m$ is the chosen generator then the corresponding homomorphism $r : C_m \to O(2s)$ sends τ to the prescribed matrix. The centralizer $C(r) \subset O(2s)$ is the product of unitary groups $U(n_1) \times \ldots \times U(n_\mu) \subset O(2s)$.

Next suppose $G = C_{nm}$. Then there is $\xi = \exp(2\pi i/mn)$. Suppose the generator $T \in G = C_{nm}$ is chosen with $T^n = \tau$. Then extend r to $R : G \to O(2s)$ by sending T to the diagonal matrix in which ξ^i appears with multiplicity n_i. Clearly $\mathrm{im}(R) \subset C(r)$ so by (35.7)

$$MSO_k(G;\ r : L \to O(2s)) \simeq MSO_k(B(G/L) \times B(U(n_1) \times \ldots \times U(n_\mu))).$$

Of course

$$B(U(n_1) \times \ldots \times U(n_\mu)) = B(U(n_1)) \times \ldots \times B(U(n_\mu))$$

and hence $H_*(B(U(n_1) \times \ldots \times U(n_\mu));Z)$ has no torsion. Thus in view of the Künneth formula (19.2) we can finally state

$$MSO_*(G,\ r : L \to O(2s)) \simeq$$
$$MSO_*(B(G/L)) \otimes MSO_*(B(U(n_1)) \otimes \ldots \otimes MSO_*(BU(n_\mu)),$$

tensor products taken over MSO_*.

We might wish to consider at once all representations of C_m which leave only o-vector fixed. In that case we simply take a direct sum over all ordered paritions $(n_1,\ldots,n_\mu)$, $n_1 + \ldots + n_\mu = s$, of s into non-negative integers.

43. Adjacent families

In this section we finally come to the case for which $\mathcal{O}_*(G;F,F')$ is computable. We assume G is a finite cyclic group of odd order. To say that F,F' are *adjacent families* means that F' is obtained by the deletion of a single (maximal) subgroup L which belongs to F.

(43.1) THEOREM: *If* $F \supset F'$ *are adjacent families of subgroups of* G *which differ only by* L *then* $\mathcal{O}_n(G;F,F') \simeq \sum_k MSO_k(G;\ L \to O(n-k))$.

This is the main objective of this section and a few explanatory comments are in order. In $MSO_k(G;\ L \to O(n-k))$ we shall only form the direct sum over those conjugacy classes of representations of L which leave nothing but o-vector fixed and therefore we are restricted to those values of k for which $n-k$ is even.

Suppose (G,W^n) represents an element of $\mathcal{O}_n(G;F,F')$. Let $F(L) \subset W^n$ be the set of stationary points under the subgroup L. Then $F(L) \cap \dot{W}^n = \emptyset$ because at $x \in \dot{W}^n$ the isotropy group G_x belongs to F' which does not contain L. Furthermore, since L is a maximal element in F, the quotient group acts freely $(G/L,F(L))$. If F^k denotes the union of the k-dimensional components of $F(L)$ then, since G is an orientation preserving group of diffeomorphisms, $F^k \neq \emptyset$ implies $n-k = 2s$. For each k with $n-k$ even let $r_1,\ldots,r_t$ denote the representatives, as described in Section 42, of those representations $r : L \to O(n-k)$ we leave only o-vector fixed. The value of t depends on k and is equal to the number of ordered partitions of s into μ non-negative integers. At each point $x \in F^k$ we shall then obtain, in the normal fibre to the component of F^k containing x, an orthogonal representation of L leaving only o fixed. This will be conjugate to exactly one of the r_j. In this fashion we obtain a decomposition

$$F^k = F_1^k \sqcup \ldots \sqcup F_t^k$$

into a disjoint union of closed submanifolds. Each F_j^k is acted upon freely by G/L. Further, using the right principal $O(n-k)$-space associated with the normal bundle to F_j^k we obtain a triple $(G,B,O(n-k))_j$ in $\mathcal{O}$ per $(G;\ r_j : L \to O(n-k))$. But then the structure of the normal bundle can be reduced to a product of unitary groups $U(n_1) \times \ldots \times U(n_\mu)$ with $2(n_1 + \ldots + n_\mu) = n-k$. In this fashion the normal bundle to F_j^k is oriented. Then the orientation of the tangent bundle to F_j^k is chosen so that the orientation of the normal bundle followed by the orientation of the tangent bundle will agree with the orientation of the tangent bundle of W^n restricted to F_j^k. Orient the final quotient $F_j^k/G/L$ appropriately, then

$$\textstyle\sum_j [(G,B,O(n-k))_j] \in \sum_j MSO_k(G;r_j : L \to O(n-k)) = MSO_k(G;L \to O(n-k))$$

is defined. The correspondence

$$[G,W^n] \to \textstyle\sum_k MSO_k(G;\ L \to O(n-k))$$

is easily seen to be a well defined homomorphism

$$\mathcal{O}_n(G;F,F') \to \sum_k MSO_k(G;\ L \to O(n-k)).$$

This is the isomorphism in question. Let us first see it is an epimorphism. Consider some triple $[G,B,O(n-k)]$ in $MSO_k(G;\ r_j : L \to O(n-k))$. Assume the intermediate quotient manifold X^k has been oriented compatibly with the final quotient manifold. Now $O(n-k)$ acts from the left on the closed unit disk D^{n-k} so we can form the associated cell bundle

$$W^n = B \times_{O(n-k)} D^{n-k}$$

over X^k. The group G acts on W^n by $g((b,v)) = ((gb,v))$. Because X^k is oriented and because the structure group of the cell bundle is reducible to a product of unitary subgroups it will follow that W^n is oriented. Now $[G,W^n] \in \mathcal{O}_n(G;F,F')$ and $[G,B,O(n-k)]$ is its image in $\sum_k MSO_k(G;\ L \to O(n-k))$.

To see the monomorphism part, we consider a (G,W^n). Surround $F = F(L,W^n) = \bigsqcup F^k$ by a G-invariant normal tube N. Then (G,N) is also (F,F')-free and in fact, by (40.4), $[G,W^n] = [G,N]$ in $\mathcal{O}_n(G;F,F')$. Suppose now $[G,W^n]$ is in the kernel of the homomorphism. Then $\sum_k [G,B_k,O(n-k)] = o$ and each $[G,B_k,O(n-k)]$ is o. That is, there are triples $(G,\mathcal{B}_k,O(n-k))$ with $(G,\dot{\mathcal{B}}_k,O(n-k)) = (G,B_k,O(n-k))$. There is then the cell bundle

$$\mathcal{N}_k = \mathcal{B}_k \times_{O(n-k)} D^{n-k}$$

over the intermediate quotient V^{k+1} which is a compact oriented manifold. There is the action of G on $\mathcal{N}_k$. The boundary of $\mathcal{N}_k$ is N_k together with the $(n-k-1)$-sphere bundle over V^{k+1}. By angle straightening $\mathcal{N}_k$ is made smooth. Every isotropy group of G on the sphere bundle over V^{k+1} is a proper subgroup of L and with $(G,N_k) \subset (G,\dot{\mathcal{N}}_k)$ we have $[G,N_k] = o \in \mathcal{O}_n(G,F,F')$ for $o \le k \le n$. Therefore $[G,W^n] = o$. ∎

Since $G = C_{mn}$ is a cyclic group of odd order and $L = C_m$ we recall from Section 42 that $r_j : L \to O(n-k)$ is determined by $(n_1,\dots,n_\mu)$ with $2(n_1 + \dots + n_\mu) = n-k$; that $C(r_j)$ is $U(n_1) \times \dots \times U(n_\mu)$ and that there is a canonical extension $R_j : G \to U(n_1) \times \dots \times U(n_\mu)$. Thus $MSO_*(G;\ r_j : L \to O(n-k)) \simeq MSO_*(B(G/L)) \otimes MSO_*(B(C(r_j)))$ and $MSO_*(B(C(r_j)))$ is a free MSO_*-module. If we write $MSO_*(B(G/L))$ as $MSO_* \oplus \widetilde{MSO}_*(B(G/L))$ then we may say $\mathcal{O}_*(G;F,F') \simeq$

$$\sum MSO_*(B(C(r_j))) \oplus \widetilde{MSO}_*(B(G/L)) \otimes (\sum MSO_*(B((C(r_j))))).$$

The first summand has no odd torsion while the second is all the odd torsion in $\mathcal{O}_*(G;F,F')$.

Let us see how the inverse isomorphism really works; that is,

$$\sum MSO_p(B(G/L)) \otimes MSO_q(B(C(r_j))) \simeq \mathcal{O}_n(G;F,F').$$

The sum is taken over all $r_j : L \to O(2s)$ with only o-vector fixed and $p + q + 2s = n$. Given a complex vector bundle $\eta \to X^q$ over a closed oriented manifold with structure group $U(n_1) \times \ldots \times U(n_\mu)$ the fact that the image of $R_j : G \to U(n_1) \times \ldots \times U(n_\mu)$ lies in the center of the structure group allows us to obtain a *fibre preserving* action $(G,D(\eta))$ on the associated closed cell bundle. Note that $\dim D(\eta) = q + 2s$ If then $(G/L,M^p)$ is a free action on a closed oriented manifold then $G \to G/L$ defines (G,M^p) with L acting trivially. There is next the diagonal action $(G, M^p \times D(\eta))$. This yields an element of $\mathcal{O}_*(G;F,F')$ and $[G/L,M^p] \otimes [\eta] \to [G, M^p \times D(\eta)]$. In particular, every element of $\mathcal{O}_n(G;F,F')$ can be expressed as a finite sum of terms of this form $[G, M^p \times D(\eta)]$.

(43.2) LEMMA: *Suppose* G *is a cyclic group of odd order, that* $K \subset G$ *is a subgroup and that* $F \supset F'$ *are adjacent families of subgroups of* K *differing by the subgroup* L. *Then the horizontal isomorphism can be chosen so that commutativity holds in*

$$\begin{array}{ccc}
\sum MSO_*(B(K/L)) \otimes MSO_*(B(C(r_j))) & \xrightarrow[\simeq]{\Theta} & \mathcal{O}_*(K;F,F') \\
\downarrow I_* \otimes \mathrm{id} & & \downarrow E_{GK} \\
\sum MSO_*(B(G/L)) \otimes MSO_*(B(C((r_j))) & \xrightarrow[\simeq]{\Theta} & \mathcal{O}_*(G;F,F')
\end{array}$$

PROOF: First $i : K/L \subset G/L$ induces the map $I : B(K/L) \to B(G/L)$ and $I_* : MSO_*(B(K/L)) \to MSO_*(B(G/L))$ is exactly the induction $E : MSO_*(K/L) \to MSO_*(G/L)$ as proved in (41.1).

Fundamentally we take each $r_j : L \to O(2s)$ and extend to $R_j : G \to C(r_j)$ with image contained in the center of $C(r_j)$. We can then use both R_j and R_j restricted to K. Let us take $[K/L,M^p]$ in $MSO_p(B(K/L)$ and $[\eta] \in MSO_q(B(C(r_j)))$. We consider $[K/L,M] \otimes [\eta]$ in $MSO_p(B(K/L)) \otimes MSO_q(B(C(r_j)))$. There are then the actions (K,M^p) and $(K,D(\eta))$; the latter given by the restriction of the homomorphism R_j to K. We would also have $(G,D(\eta))$ and $(K,D(\eta)) = R_{KG}(G,D(\eta))$.

The composition $E_{GK}\Theta$ will send $[K/L,M] \otimes [\eta]$ to $E_{GK}([K,M] \times R_{KG}[G,D(\eta)])$ which is $E_{GK}([K,M]) \cdot [G,D(\eta)]$ by (40.6). Using (41.1), on the other hand, I_* sends $[K/L,M]$ into $E[K/L,M] = [G,N]$ where

$$N = G/L \times_{K/L} M = G \times_K M.$$

Thus $\Theta(I_* \otimes id)$ carries $[K/L,M] \times [\eta]$ into $(E_{GK}[K,M]\cdot[G,D(\eta)])$. Thus the lemma follows. ∎

(43.3) LEMMA: *Suppose* G *is a finite cyclic group of odd order,* $K \subset G$ *is a subgroup and* $F \supset F'$ *are adjacent families of subgroups of* K *which differ by* $L \subset K$. *Then*

$$E_{GK} : \mathcal{O}_n(K,F,F') \to \mathcal{O}_n(G;F,F')$$

is a monomorphism for n *odd and an isomorphism for* n *even.*

PROOF: Since $MSO_*(B(C(r_j))$ is a free MSO_*-module it will, in view of (43.2) be enough to note

$$I_* : MSO_k(B(K/L)) \to MSO_k(B(G/L))$$

is a monomorphism for n odd and an isomorphism for n even. This was shown in (41.3). ∎ It should be emphasized that $MSO_*(B(C(r_j))$ is a free module on even dimensional generators.

The final result of this section is the key to the applications.

(43.4) THEOREM: *Suppose that* G *is a finite cyclic group of odd prime power order; that* $K \subset G$ *is a subgroup and that* $F \supset F'$ *are families in* K. *Then*

$$E_{GK} : \mathcal{O}_n(K;F,F') \to \mathcal{O}_n(G;F,F')$$

is a monomorphism if n *is odd and an epimorphism if* n *is even.*

PROOF: Notice that the families are not assumed adjacent. Suppose K has order p^k. Then the families of subgroups of K are in ascending order

$$\emptyset = F_{-1} \subset F_o \subset F_1 \subset \dots \subset F_k$$

where F_i consists of all subgroups whose order divides p^i. Each consecutive pair is a pair of adjacent families. Now we study

$$E_{GK} : \mathcal{O}_n(K;F_i,F_j) \to \mathcal{O}_n(G;F_i,F_j) \qquad i > j$$

by induction on the difference $i - j$. For $i - j = 1$ the conclusion of (43.4) is just (43.3). Suppose (43.4) has been shown for all $i - j < m$ where $m \geq 2$. Consider then $i - j = m \geq 2$. There is then the triple $F_i \supset F_{i-1} \supset F_j$ and the diagram

$$\begin{array}{ccccccccc}
\mathcal{O}_{n+1}(K;F_i,F_{i-1}) & \to & \mathcal{O}_n(K;F_{i-1},F_j) & \to & \mathcal{O}_n(K;F_i,F_j) & \to & \mathcal{O}_n(K;F_i,F_{i-1}) & \to & \mathcal{O}_{n-1}(K;F_{i-1},F_j) \\
\downarrow E^1_{GK} & & \downarrow E^2_{GK} & & \downarrow E^3_{GK} & & \downarrow E^4_{GK} & & \downarrow E^5_{GK} \\
\mathcal{O}_{n+1}(G;F_i,F_{i-1}) & \to & \mathcal{O}_n(G;F_{i-1},F_j) & \to & \mathcal{O}_n(G;F_i,F_j) & \to & \mathcal{O}_n(G;F_i,F_{i-1}) & \to & \mathcal{O}_{n-1}(G;F_{i-1},F_j)
\end{array}$$

If n is odd then E^2_{GK} and E^4_{GK} are monomorphisms by induction and (43.3) while E^1_{GK} is an isomorphism by (43.3). From the five-lemma it then follows E^3_{GK} is a monomorphism for n odd. Next if n is even then E^2_{GK} is an epimorphism by induction while E^4_{GK} is an isomorphism by (43.3). Furthermore E^5_{GK} is a monomorphism by induction. Again the five-lemma shows E^3_{GK} is an epimorphism for n even. ∎

44. Applications

We shall present in this section two applications of the foregoing to orientation preserving diffeomorphisms of odd prime power period on closed oriented manifolds.

(44.1) THEOREM: *If M^n is a closed oriented manifold which admits an orientation preserving diffeomorphism of odd prime power period, p^k, having no fixed points (stationary points) then the oriented bordism class of M^n is divisible by* p.

PROOF: If n is odd there is nothing to prove since MSO_n consists entirely of 2-torsion if n is odd. Then let (G,M^n) denote an action of the cyclic group of order p^k on a closed oriented even dimensional manifold as a group of orientation preserving diffeomorphisms. Suppose no point is fixed under the entire group and let F be the family of all proper subgroups of G. Then $[G,M^n] \in \mathcal{O}_n(G;F)$. Now if $K \subset G$ is the subgroup of order p^{k-1} then F is actually the family of all subgroups of K. We apply (43.4), with n even, so that

$$E_{GK} : \mathcal{O}_n(K;F) \to \mathcal{O}_n(G;F)$$

is an epimorphism. Then there is a smooth orientation preserving action on a closed manifold (K,V^n) with $E_{GK}[K,V^n] = [G,M^n]$. In particular, $[M^n] = [G \times_K V^n]$ in MSO_n. But $G \times_K V^n$ is a fibre bundle over G/K so that $G \times_K V^n$ is actually p disjoint copies of V^n. Thus clearly $p[V^n] = [M^n]$. ∎

A comment is in order. For any finite group G an ideal $I(G) \subset MSO_*$ can be described as those oriented bordism classes which admit a representative that can be acted on by G as a group of orientation preserving diffeomorphisms without any stationary points. For G a cylic group of odd prime power order we have just shown that this ideal is $pMSO_*$. For G an elementary abelian p-group, p still odd, the structure of this ideal was first demonstrated by Floyd [F] and then this was utilized and expanded upon by Tammo tom Dieck [D_2].

The next theorem was conjectured in the first edition. It was first verified by Atiyah and Bott using the original version of their fixed point formula [AB].

(44.2) THEOREM: *An orientation preserving diffeomorphism of odd prime power period on a closed oriented positive dimensional manifold cannot have exactly one stationary point.*

PROOF: We consider (T,M^n) of odd prime power period p^k. We shall prove the result by induction over k. Suppose it has been shown for period p^i, $i < k$ ($i = o$ is trivial of course). Suppose (T,M^n) has period p^k with exactly one fixed point, x. Clearly n must be even. We claim now that T generates a free action of the cyclic group of order p^k on a deleted invariant neighborhood of x. If $k = 1$ this is obvious. Otherwise, introduce $t = T^{p^{k-1}}$ which has period p. Let W be the component of the fixed point set F(t) which contains x. Certainly W is a closed connected oriented submanifold which is T-invariant. The restriction of T to W still has exactly one fixed point, x, but now its period is p^{k-1} so by induction W is the single point x. Hence if (G,M^n) is the cyclic group of order p^k generated by T then at this stationary point x we receive an (orthogonal) representation (G,R^n) on the tangent space at x in which G acts freely on the unit sphere. Call this (G,S^{n-1}), then $[G,S^{n-1}] \in MSO_{n-1}(G)$. On the complement $M^n \setminus V$ of a small open invariant cell around x the group G will act with only proper subgroups as isotropy groups since x is the only stationary point. We take it that the boundary of $(G, M \setminus V)$ is (G,S^{n-1}).

Now if F is the family of all proper subgroups of G and F' consists of the trivial subgroup then $[G,S^{n-1}] \in \mathcal{O}_{n-1}(G;F')$ and it is sent to zero by

$$\mathcal{O}_{n-1}(G;F') \to \mathcal{O}_{n-1}(G;F).$$

Again $K \subset G$ is the subgroup of order p^{k-1} so that F is also the family of all subgroups of K. Recall $n = 2k$ and consider

$$\begin{array}{ccccc} \mathcal{O}_{2k}(K;F,F') & \xrightarrow{\partial} & \mathcal{O}_{2k-1}(K;F) & \xrightarrow{\alpha} & \mathcal{O}_{2k-1}(K;F) \\ \downarrow E^1_{GK} & & \downarrow E^2_{GK} & & \downarrow E^3_{GK} \\ \mathcal{O}_{2k}(G;F,F') & \xrightarrow{\partial'} & \mathcal{O}_{2k-1}(G,F') & \xrightarrow{\alpha'} & \mathcal{O}_{2k-1}(G;F) \end{array}$$

Then E^1_{GK} is an epimorphism while E^2_{GK} and E^3_{GK} are monomorphisms by (43.4). Since $X = [G,S^{2k-1}]$ in $\mathcal{O}_{2k-1}(G;F')$ lies in the kernel of α' we can write $X = \partial'Y$. Since E^1_{GK} is an epimorphism there is a $Z \in \mathcal{O}_{2k}(K;F,F')$ with $E^1_{GK}(Z) = Y$. Hence $E^2_{GK}(\partial Z) = X$. That is, $[G,S^{2k-1}] \in MSO_{2k-1}(G)$ lies in the image of

$$E_{GK} : MSO_{2k-1}(K) \to MSO_{2k-1}(G).$$

But then by (41.1), $\mu[G,S^{2k-1}]$ lies in the image of $i_{GK} : H_{2k-1}(K;Z) \to H_{2k-1}(G;Z)$. Now i_{GK} has image the subgroup of index p and this contradicts the fact that $\mu[G,S^{2k-1}]$ generates $H_{2k-1}(G;Z)$. ∎

We caution the reader that the result is invalid for cyclic groups of odd composite order, [CF_3].

45. Some examples

To discuss the work of tom Dieck, even as restricted to maps of odd prime power period, it is first necessary to construct some explicit periodic maps. We fix an odd prime power period p^k. To begin with we want to give on each complex projective space a map of period p^k for which the set of fixed points has the lowest possible dimension.

(45.1) LEMMA: *If* $n \geq 0$ *and* $0 \leq j < p^k$ *then there is a map* $(\tau, CP(np^k + j))$ *of period* p^k *for which* $\dim F(\tau) \leq 2n$.

PROOF: This means no component of the fixed point set has dimension (real) larger than 2n. We think of a point in $CP(np^k + j)$ as expressed in the form

$$[Z_1,\dots,Z_n; z_{np^k+1},\dots, z_{np^k+j+1}]$$

with $Z_i = (z_{i,1},\dots,z_{i,p^k})$. Let $\lambda = \exp(2\pi i/p^k)$. By λZ_i we mean $(z_{i,1}, \lambda z_{i,2},\dots, \lambda^{p^k-1} z_{i,p^k})$. Then

$$\tau[Z_1,\ldots,Z_n;\ z_{np^k+1},\ldots,z_{np^k+j+1}] =$$

$$[\lambda Z_1,\ldots,\lambda Z_n;\ z_{p^k+1},\lambda z_{p^k n+2},\ldots,\lambda^j z_{p^k n+j+1}]$$

To see a component $F_a \subset F(\tau)$ with highest possible dimension 2n selection integer a, $1 \leq a \leq j+1$, and put all co-ordinates equal to zero except z_{np^k+a} and $\left\{z_{i,a}\right\}_{i=1}^{n}$. All components of $F(\tau)$ other than these have dimension $2n - 2$. ∎

Let us turn next to the hyper-surface $H(r,s) \subset CP(r) \times CP(s)$ of type (1,1). We always take it that $r \leq s$. Then $H(r,s)$ consists of all pairs of points whose homogeneous coordinates satisfy $z_1\hat{z}_1 + \ldots + z_{r+1}\hat{z}_{r+1} = o$. Now there are the maps $(\tau, CP(r))$ and $(\hat{\tau}, CP(s))$ as constructed above and if we take on $CP(r) \times CP(s)$ the diagonal map defined by τ and $\hat{\tau}^{-1}$ then $H(r,s)$ is invariant and there results a map $(T, H(r,s))$ of period p^k. The dimension of $H(r,s)$ is $2(r+s-1)$ and we want to investigate $\dim F(T)$.

(45.2) LEMMA: *If* $r = s = o$ *(mod* p^k*) then* $\dim F(T) \leq \dim F(\tau) + \dim F(\hat{\tau}) - 2$ *Otherwise* $\dim F(T) \leq \dim F(\tau) + \dim F(\hat{\tau})$.

PROOF: Write $r = np^k + j$ and $s = mp^k + q$. If either j or q is positive then it is possible to choose *distinct* integers $1 \leq a \leq j+1$, $1 \leq b \leq q+1$. There are then the components $F_a \subset F(\tau)$ and $\hat{F}_b \subset F(\hat{\tau})$ with dimensions 2n and 2m respectively. But obviously the product $F_a \times \hat{F}_b$ lies in $H(r,s)$ and hence in $F(T)$. If $j = q = o$, however, we will only have $F_1 \subset F(\tau)$ and $\hat{F}_1 \subset F(\hat{\tau})$ so that in effect we only receive an $H(n,m)$ in $F(T)$, thereby losing two dimensions. ∎

Recall that Milnor determined that MSO_*/Tor is a graded polynomial ring over Z with a generator in each dimension 4k.

(45.3) LEMMA: *For each dimension* 4i *there is a map of period* p^k *on a closed oriented manifold* (T, M^{4i}) *for which* $[M^{4i}]$ *is a generator in* MSO_*/Tor *and* $\dim F(T) \leq 2[2i/p^k]$.

PROOF: A generator in dimension 4i can be obtained as an integral linear combination of $[CP(2i)]$ and the $[H(r,s)]$ with $r+s-1 = 2i$ [St_2]. Write $2i = N_p k + J$, $o \leq J \leq p^k - 1$ so that $N = [2i/p^k]$. Of course we know for $(\tau, CP(2i))$ that $\dim F(\tau) \leq 2N$. Next consider $r \leq s$ with $r + s - 1 = zi$ and $r = np^k + j$, $s = mp^k + q$. Then $r + s = (n+m)p^k + j + q = Np^k + J + 1$. As long as $J < p^k - 1$ then $N = [(r+s)/p^k] = n + m$ or $n + m + 1$. But

$\dim F(T) \leq 2(n+m) \leq 2N$ by (45.2). Now if $J = p^k - 1$ then $(N+1)p^k = r + s = (n+m)p^k + j + q$. Thus either $j = q = o$ and $n + m = N+1$ or $j + q = p^k$ and $n+m = N$. In the second case there is no problem as $\dim F(T) \leq 2(n+m) \leq 2N$, while in the first case we apply the first part of (45.2) to find $\dim F(T) \leq 2N$. ■

46. Some results of tom Dieck

Let now G be the cyclic group of order p^k. Let $F \supset F'$ be the adjacent pair, differing by G, which is the family of all subgroups followed by the family of all proper subgroups. Since G = L in this case we can, from Section 41, say that

$$\mathcal{O}_n(G;F,F') \simeq \sum MSO_q(B(C(r_j)))$$

where the sum as usual is run over all $r_j : G \to O(2s)$ leaving only o-vector fixed and $q + 2s = n$. It is also possible to describe $\mathcal{O}_*(G;F,F')$ is a fashion analogous to the algebra M_* introduced in Section 25 in connection with the unoriented case.

Put $\mu = (p^k - 1)/2$, then if $(n_1,\ldots,n_\mu)$ is an ordered μ-tuple of non-negative integers then an element of $MSO_q(B(U(n_1)) \times \ldots \times B(U(n_\mu)))$ can be thought of as an ordered μ-tuple of unitary bundles over a closed oriented manifold. This would look like $[(\xi_1,\ldots,\xi_\mu) \to X^q]$. The complex dimension of ξ_j is n_j. The total degree of this object is $n = q + 2(n_1 + \ldots + n_\mu)$. We do allow o-bundles. Take M_n to consist of all these bordism classes having total degree n. Then $\mathcal{O}_n(G;F,F') \simeq M_n$. Then $MSO_n \subset M_n$ as a closed manifold with all ξ_j of dimension o. Addition is by disjoint union. A product in $M_* = \sum M_n$ is given as follows. For two objects $(\xi_1,\ldots,\xi_\mu) \to X^q$ and $(\eta_1,\ldots,\eta_\mu) \to Y^p$ we form $(\xi_1 \times \eta_1,\ldots,\xi_\mu \times \eta_\mu) \to X^q \times Y^p$ where $\xi_j \times \eta_j$ is the external Whitney sum. This is commutative. In fact, since $MSO_* \subset M_*$ we can think of M_* as a graded commutative algebra with identity over MSO_*.

Actually of course M_* is

$$MSO_*(B(U(n_1))) \otimes \ldots \otimes MSO_*(B(U(n_\mu)))$$

where the sum is run over all sequences of non-negative integers $(n_1,\ldots,n_\mu)$. Hence M_* is a polynomial algebra over MSO_*. Generators may be described as follows. Consider a class $c(k,j)$ for each ordered pair of integers $k \geq o$ and $1 \leq j \leq \mu$ which is represented by

$(\eta_1,\dots,\eta_\mu) \to CP(k)$ where η_i is the o-bundle if $i \neq j$ and $\eta_j \to CP(k)$ is the complex Hopf line bundle. The total degree of $c(k,j)$ is $2(k+1)$. Note $c(o,j)$ is included, $1 \leq j \leq \mu$ and do not confuse these with $[CP(o)] = 1 \in MSO_o$.

Remember that MSO_*/Tor is a polynomial ring over Z, so that M_*/Tor may also be thought of as a polynomial ring over Z. Take generators for MSO_*/Tor together with the $c(k,j)$.

We shall be concerned with certain subrings of M_*/Tor. For any $m \geq o$ let $F(m) \subset M_*/Tor$ be the subring which, modulo torsion, is generated by all $[(\xi_1,\dots,\xi_\mu) \to X^q]$ with $q \leq m$. Now $F(m)$ is plainly a polynomial ring over Z of finite transcendence degree. All $c(k,j)$ with $2k \leq m$ belong to $F(m)$ and for $4n \leq m$ we get a generator out of MSO_{4n}.

(46.1) LEMMA: *For* $m \geq o$ *the transcendence degree of* $F(m)$ *over* Z *is* $[m/4] + \mu(1 + [m/2])$. ■

In other words, $F(m) \subset M_*/Tor$ is a polynomial ring over Z on this many generators.

Let us return to $\mathcal{O}_*(G) = \mathcal{O}_*(G;F)$. There is an augmentation homomorphism $\varepsilon : \mathcal{O}_*(G) \to MSO_*$ which is simply $[G,M] \to [M]$. Now the image of $\mathcal{O}_*(G;F') \to \mathcal{O}_*(G)$ is an ideal J. According to (44.1) the augmentation will induce $\mathcal{O}_*(G)/J \to MSO_*/pMSO_*$. Furthermore there is a multiplicative monomorphism

$$o \to \mathcal{O}_*(G)/J \to M_*$$

arising from the exact triangle

$$\begin{array}{ccc} \mathcal{O}_*(G;F') & \longrightarrow & \mathcal{O}_*(G;F) \\ & \nwarrow \quad \swarrow & \\ & \mathcal{O}_*(G;F,F'). & \end{array}$$

To simplify the computations let us assume that $m = 4s$ so that $F(m)$ has degree $s + \mu(1 + 2s) = sp^k + (p^k - 1)/2$ over Z. According to (45.3) choose, for $1 \leq n \leq sp^k(p^k - 1)/2$ a map (T,M^{4n}) of period p^k so that $[M^{4n}]$ is a generator of MSO_*/Tor and $\dim F(T) \leq 2[2n/p^k]$. In particular with $n = sp^k + (p^k - 1)/2$, $2[2n/p^k] = 4s = m$. Thus take $R(m) \subset \mathcal{O}_*(G)/J$ to be the subring with identity generated by these classes $[T,M^{4n}]$, $1 \leq n \leq sp^k + (p^k - 1)/2$. Under $M_*(G)/J \to M_*/Tor$ the ring $R(m)$ is carried into $F(m)$.

(46.2) LEMMA: *The image of* R(m) *in* F(m) *has the same transcendence degree over* Z *as does* F(m).

PROOF: For the degree to decrease there would have to be a non-trivial polynomial with integral coefficients in the classes $[T, M^{4n}]$ which goes to o in M_*/Tor. We can assume the g.c.d. of the coeffficients of this polynomial is 1. So multiplying this polynomial by 2 if necessary we can assume it belongs to J. Now apply

$$\mathcal{O}_*(G) \to \mathcal{O}_*(G)/J \to MSO_*/pMSO_*.$$

The polynomial is sent to o, but this obviously contradicts the fact that the $[M^{4n}]$ are polynomial ring generators for MSO_*/Tor and hence for $MSO_*/pMSO_*$ as a polynomial ring over Z/pZ. ∎

This shows every element of the polynomial ring F(m) is algebraic over R(m).

If m = 4s put $t(m) = p^k s + (p^k - 1)/2$, the common transcendence degree. We obtain

(46.3) TOM DIECK: *If a closed oriented manifold admits a map of period* p^k *for which no component of the fixed point set has dimension exceeding* m = 4s *then in* $MSO_*/pMSO_*$ *the bordism class of this manifold lies in the subring generated by* $\sum_0^{t(m)} MSO_{4n}/pMSO_{4n}$.

PROOF: Suppose (G, X^s) is an action with $\dim F(G) \leq m$. Then under $\mathcal{O}_*(G) \to M_*/\mathrm{Tor}$ this class will be carried into F(m). We can assume that this image is not zero for otherwise $2[G, X^s] \in J$ which would imply that $[X^s] \in pMSO_*$.

We want to apply (46.2). We can find polynomials with integral coefficients

$$q_j = q_j([T, M^4], \ldots, [T, M^{4t(m)}]) \quad o \leq j \leq r$$

so that q_0 is non-trivial; some coefficient of some q_j is not divisible by p and

$$\sum_0^r q_j \cdot [G, X^s]^j$$

lies in J. We apply $\mathcal{O}_*(G)/J \to MSO_*/pMSO_*$ and we obtain a relation (non-trivial)

$$\sum_0^r q_j([M^4], \ldots, [M^{4t/m}]) \cdot [X^s]^j = o.$$

Now $MSO_*/pMSO_*$ is a polynomial ring over Z/pZ and these $[M^{4i}]$ are taken to be among its generators. Therefore $[X^s]$ in $MSO_*/pMSO_*$ lies in the subring generated by $\sum_0^{t(m)} MSO_{4n}/pMSO_{4n}$. ∎

For example, if $m = o$ then for any action (G,X^s) having a finite number of stationary points the oriented bordism class $[X^s] \in MSO_s/pMSO_s$ will lie in the subring generated by

$$\sum_0^{(p^k-1/2)} MSO_{4n}/pMSO_{4n}.$$

Now we have outlined an example of tom Dieck's work to generally indicate the idea. Actually, of course, the results are far more general and we do refer the reader now to $[D_1-D_5]$.

47. Manifolds with all Pontrjagin numbers divisible by p

This is only a discussion section. Although proofs were included in the first edition of this monograph a more detailed treatment is now available in $[St_2]$. If p is a fixed odd prime it may be asked if the divisibility of all Pontrjagin numbers of a closed oriented manifold by p always implies that the oriented bordism class $[M^{4n}]$ is also divisible by p in MSO_{4n}. As we remarked earlier this is indeed the case if $n < p - 1/2$. However, the most elementary counter example to the general question is provided by CP(2). Here there is only one Pontrjagin number and it is divisible by 3. Yet the signature of the manifold is 1 so [CP(2)] is obviously not divisible by 3. To pursue this further, it is not difficult to see that for any odd prime all the Pontrjagin numbers of CP(p - 1) are divisible by p while each of these manifolds has signature 1.

Thus it is apparent that we must find a different question. For each odd prime define an ideal $I(p) \subset MSO_*$ by $[M^n] \in I_n(p)$ if and only if all Pontrjagin numbers of $[M^n]$ are divisible by p. Clearly $I_n(p) = MSO_n$ if $n \not\equiv o \pmod 4$, and $I(p) \supset pMSO$. We ask now for the structure of the ideal I(p). Recall again that for each $k > o$ Milnor showed that there is a closed oriented manifold Y^{4k} for which $s_k(Y) = 1$ if $2k + 1$ is not a prime power or $s_k(Y) = p$ if $2k+1 = p^j$, p a prime. We shall call such manifolds *Milnor base elements for* MSO_*/Tor.

(47.1) THEOREM: *For each odd prime* p *there are Milnor base elements* Y^{2p^j-2}, $1 \leq j < \infty$, *all of whose Pontrjagin numbers are divisible by* p. *Furthermore,* I(p) *is the ideal generated by* Y^o, *the manifold with* p *points together with these* Y^{2p^j-2}. ∎

We can say that $MSO_{4k}/I_{4k}(p)$ is an elementary abelian p group whose rank, d(k), is the number of partitions of k into non-negative integers none of which have the form $(p^j - 1)/2$. Of course then $I_{4k}(p)/pMSO_{4k}$ is also an elementary abelian p-group whose rank, d'(k), is the number of partitions of k into non-negative integers at least one of which has the form $(p^j - 1)/2$.

The ideal I(p), as well as the special manifolds Y^{2p^j-2}, play a key role in the study via bordism of actions of abelian p-groups on closed oriented manifolds. We shall at least begin to see this by further consideration of $MSO_*(C_p)$.

We would add here that it is possible to detect the divisibility of an oriented bordism class by introducing K-theory characteristic numbers. This question was completely settled by work of Stong [St_2], and Hattori.

For later reference we close this section with the following lemma.

(47.2) LEMMA: *For a complex* X *let* $\alpha_1,\dots,\alpha_r$ *be homogeneous bordism classes in* $MSO_*(X)$ *and suppose that*

$$MSO_*(X) \xrightarrow{\mu} H_*(X;Z) \longrightarrow H_*(X;Z/pZ)$$

maps $\alpha_1,\dots,\alpha_r$ *into linearly independent elements of* $H_*(X;Z/pZ)$. *If* $[M^{n_1}],\dots,[M^{n_r}]$ *are elements of* MSO_* *for which* $\sum \alpha_k[M^{n_k}] = o$ *in* $MSO_*(X)$ *then all the* $[M^{n_k}]$ *lie in* I(p).

PROOF: Suppose that α_k is represented by a map $f_k : V^{m_k} \to X$. There is the projection $\pi_k : V^{m_k} \times M^{n_k} \to V^{m_k}$. By hypothesis

$$\sum [V^{m_k} \times M^{n_k}, f_k\pi_k] = o$$

in $MSO_*(X)$. We may just as well assume that $m_k + n_k$ is constant independent of k. Just as in Section 17, it will follow that if $c \in H^*(X;Z/pZ)$ and if p_ω denotes a product of Pontrjagin classes taken modulo p than in Z/pZ $\sum_k \langle p_\omega(V^{m_k} \times M^{n_k})\pi_k^* f_k^*(c), \sigma(V^{m_k} \times M^{n_k})\cdot\rangle = o$. Here $\sigma(V^{m_k} \times M^{n_k})$ is the product orientation.

Suppose $m_1 \geq \dots \geq m_r$, then we can argue by induction. That is, assume $M^{n_1},\dots,M^{n_{k-1}}$ have all their Pontrjagin numbers divisible by p. Note $n_1 \leq \dots \leq n_{k-1}$. There is a cohomology class $c \in H^{m_k}(X;Z/pZ)$ with $\langle c,\mu_p(\alpha_k)\rangle = 1$ and $\langle c,\mu_p(\alpha_j)\rangle = o$ in Z/pZ for $j \neq k$. Since p is odd we have the Whitney sum theorem for Pontrjagin classes

$$p_\omega(V^{m_j} \times M^{n_j}) = 1 \otimes p_\omega(M^{n_j}) + \sum a_s \otimes b_t$$

with $\deg(a_s) > 0$. Finally, $\pi_j^* f_j^*(c) = f_j^*(c) \otimes 1$ and the b_t are products of mod p Pontrjagin classes of M^{n_j}. Now then $\langle p_\omega \cdot \pi_j^* f_j^*(c), \sigma\rangle =$ $\langle f_j^*(c) \otimes p_\omega(M^{n_j}), \sigma\rangle + \langle \sum a'_s \otimes b_t, \sigma\rangle$ with $\deg(a'_s) > m_k$. But $\sigma(V^{m_j} \times M^{n_j}) = \sigma(V^{m_j}) \times \sigma(M^{n_j})$ and therefore

$$\langle p_\omega \cdot \pi_j^* f_j^*(c), \sigma\rangle = \langle f_j^*(c), \sigma(V^{m_j})\rangle\langle p_\omega(M^{n_j}), \sigma(M^{n_j})\rangle$$
$$+ \sum \langle a'_s, \sigma(V^{m_j})\rangle\langle b_t, \sigma(M^{n_j})\rangle.$$

For $j < k$ the above is o by induction. If $j > k$ then $\langle f_j^*(c), \sigma(V^{n_j})\rangle = \langle c, \mu_p(\alpha_j)\rangle = 0$ while $\langle a'_s, \sigma(V^{m_j})\rangle = 0$ because $\deg(a'_s) > m_j$. Thus all we have left is

$$\begin{aligned} o &= \sum \langle p_\omega \cdot \pi_j^* f_j^*(c), \sigma\rangle \\ &= \langle p_\omega \cdot \pi_k^* f_k^*(c), \sigma\rangle \\ &= \langle c, \mu_p(\alpha_k)\rangle\langle p_\omega(M^{n_k}), \sigma(M^{n_k})\rangle \\ &= \langle p_\omega(M^{n_k}), \sigma(M^{n_k})\rangle \end{aligned}$$

in Z/pZ. This completes the lemma. ■

48. Fixed point sets with trivial normal bundles

We return now to an orientation preserving diffeomorphism of odd prime period p on a closed oriented manifold. There is $F \subset M^n$, the set of fixed points and we can write $F = \bigsqcup F^k$ where $n - k$ is even and F^k is the union of the k-dimensional components of the fixed point set. If F_c^k is such a non-empty component then the normal bundle $\eta \to F_c^k$ is given the structure of a $U(n_1) \times \ldots \times U(n_\mu)$ bundle with $2(n_1 + \ldots + n_\mu) = n - k$. *In this section we shall always assume the numbers* $n_1, \ldots, n_\mu$ *depend only on* k *and not on the particular component of* F^k. *Furthermore,* $F^n = \emptyset$ *and for each component* F_c^k *we assume that* $\eta \to F_c^k$ *is bordant in* $MSO_k(B(U(n_1)) \times \ldots \times B(U(n_\mu))$ *to the product bundle over* F_c^k.

To abbreviate all this we shall simply say that (T, M^n) has a *fixed point set with trivial normal bundles*. As the reader will guess, this is the simplest case for us to handle and there are some interesting examples.

(48.1) THEOREM: *If* $T : M^n \to M^n$ *is an orientation preserving diffeomorphism of odd prime period on a closed oriented manifold for which the fixed point set has trivial normal bundles then* $[M^n] \in I(p)$ *and* $[F^k] \in I(p)$ *for* $0 \leq k < n$.

PROOF: Let us take stock of our situation. We have

$$\to \mathcal{O}_n(C_p) \longrightarrow \mathcal{M}_n \xrightarrow{\partial} MSO_{n-1}(C_p) \to$$

where $\mathcal{M}_n = \sum MSO_k(B(U(n_1)) \times \ldots \times B(U(n_\mu))$.

The first homomorphism assigns the normal data to the fixed point set. By hypothesis, (T,M^n) has a fixed point set with trivial normal bundles so that $F^n = \emptyset$ and $[T,M^n] \to \sum [C_p,D^{n-k}][F^k]$ where the sum extends over those k with $o \leq k < n$ and $n - k$ even, say $n - k = 2m$. We recall that $(n_1,\ldots,n_\mu)$ defines a unitary representation of C_p on C^m and (C_p,D^{n-k}) is the restriction of this action to the closed unit cell. In terms of the $c(k,j)$ from Section 46, $[C_p,D^{n-k}] = c(o,1)^{n_1}\ldots c(o,\mu)^{n_\mu}$.

The induced action on $\dot{D}^{n-k}$ is free and by exactness if we apply ∂ we find

$$\sum [C_p,S^{n-k-1}][F^k] = o$$

in $MSO_{n-1}(C_p)$. Since $n - k - 1 \geq 1$ we see the homology classes $\mu_p[C_p,S^{n-k-1}]$ are linearly independent in $H_*(C_p;Z/pZ)$ so that we can just apply (47.2) directly to see that for each k, $[F^k] \in I(p)$.

To see that $[M^n]$ belongs to I(p) we must use (37.2). This tells us how $[M^n]$ is determined modulo $pMSO_n$ by the normal data to the set of fixed points. In the case at hand we get

$$\sum [C_p,S^{n-k+1}][F^k] - [C_p,S^1][M^n]$$

$= o$ in $MSO_{n+1}(C_p)$. Here the action of C_p on S^{n-k+1} is obtained from that on S^{n-k-1} by adding another complex coordinate on which C_p acts as multiplication by $\exp(2\pi i/p)$. In other words, change $(n_1,\ldots,n_\mu)$ to $(n_1 + 1, n_2,\ldots,n_\mu)$. We apply (47.2) to complete the argument. ∎

To proceed we need a lemma.

(48.2) LEMMA: *If* $[T,X^{2n-1}]$ *is an element of* $MSO_{2n-1}(C_p)$ *for which* $\mu[T,X^{2n-1}] \neq o$ *in* $H_{2n-1}(C_p;Z)$ *then the ideal of elements in* MSO_* *which annihilate* $[T,X^{2n-1}]$ *is* $p^{a+1}MSO_*$, $a(2p-2) < 2n - 1 < (a+1)(2p-2)$

PROOF: We proceed by induction on n. It follows for $n = 1$ by (47.2, 36,7, 36.8). In general from (38.1) it will follow that $p^{a+1}MSO_*$ lies in this annihilator ideal. Suppose the lemma is proved for $2n-3$. If $[T,X^{2n-1}][M^m] = o$ then surely $(\Delta_1[T,X^{2n-1}])([M^m])$ is also o, and $\mu(\Delta_1[T,X^{2n-1}]) \neq o$. Hence if $a(2p-2) + 3 \leq 2n - 1 < (a+1)(2p-2)$ then

$[M^m]$ lies in $p^{a+1}MSO_*$ by induction and so the assertion follows for $2n-1$.

If, on the other hand, $2n-1 = a(2p-2)+1$ then by induction $[M^m] \in p^a MSO_m$. Let us write $p^a[V^m] = [M^m]$. Then $p^a[T, X^{2n-1}][V^m] = o$. But then by (38.2) we can write

$$b[T_1, S^1][CP(p-1)]^a[V^m] = o$$

where $b \not\equiv o \pmod p$. Inductively this implies $b[CP(p-1)]^a[V^m] \in pMSO_*$. However, $[CP(p-1)]$ gives one of the generators of $MSO_*/pMSO_*$ as a polynomial algebra over Z/pZ. Therefore $[V^m] \in pMSO_m$ and $[M^m] \in p^{a+1}MSO_m$. ∎

We can tackle (48.1) again, this time assuming that all non-empty components of the fixed point set have the same dimension.

(48.3) THEOREM: *If* (T, M^n) *has a fixed point set with trivial normal bundles and if all the components of* F *have the same dimension,* m, *then* $[F^m] \in p^{a+1}MSO_m$ *where* $(a-1)(2p-2) < n-m \le a(2p-2)$. *If* $n-m \not\equiv a(2p-2)$ *then* $[M^n] \in pMSO_n$ *while if* $n-m = a(2p-2)$ *then*

$$[M^n] = b[CP(p-1)]^a[X^m]$$

in $MSO_n/pMSO_n$ *where* $b \not\equiv o \pmod p$ *and* $p^a[X^m] = [F^m]$.

PROOF: Under these hypotheses we have simply $[C_p, S^{n-m-1}][F^m] = o$ in $\widetilde{M}SO_{n-1}(C_p)$. Thus by (48.2), $[F^m] \in p^a MSO_m$ where $(a-1)(2p-2) < n-m \le a(2p-2)$. Also from (37.2)

$$[T, S^{n-m-1}][F^m] = [T, S^1][M^n] = p^a[T, S^{n-m+1}][X^m].$$

If $n-m \not\equiv a(2p-2)$ then $p^a[T, S^{n-m+1}] = o$ and so $[M^n] \in pMSO_n$ by (48.2). If $n-m = a(2p-2)$ then by (38.2)

$$p^a[T_1, S^1][X^m] = [T_1, S^1](b[CP(p-1)]^a[X^m]) = [T_1, S^1][M^n].$$

This completes the proof. ∎

The assumption of trivial normal bundles is crucial here. For example, consider $CP(n)$ with $1 \le n < p-1$. Define a map of period p on $CP(n)$ by

$$T[z_1, \ldots, z_{n+1}] = [z_1, \lambda z_2, \ldots, \lambda^n z_{n+1}].$$

The fixed point set consists of $n+1$ isolated points, yet $CP(n)$ does not lie in $pMSO_{2n}$ for n even. We cannot use (48.3) because the numbers $(n_1, \ldots, n_\mu)$ are not independent of the fixed point.

49. Maps on Milnor manifolds

In this section we shall construct some closed oriented manifolds Y^{2p^j-2} all of whose Pontrjagin numbers are divisible by p and for which $s_{\frac{p^j-1}{2}}(Y) \equiv p \pmod{p^2}$. In the next section we then shall see how such manifolds play a role in the module structure of $\widetilde{M}SO_*(C_p)$. These are *Milnor base elements for* $MSO_*/pMSO_*$.

First we recall some results of Borel and Hirzebruch [BH]. Suppose $\xi \to X$ is a complex n-plane bundle with structure group U(n). The center of U(n) is $U(1) = S^1$ embedded as scalar matrices. If $S(\xi) \to X$ is the associated (2n-1)-sphere bundle, then S^1 acts freely on $S(\xi)$ sending each fibre into itself. The resulting quotient $S(\xi)/S^1 = CP(\xi) \to X$ is the associated complex projective space bundle with CP(n - 1) as fibre. The quotient $S(\xi) \to CP(\xi)$ is a principal U(1) and so has a Chern class $a \in H^2(CP(\xi);Z)$. Moreover, if $\pi: CP(\xi) \to X$ denotes the projection map then every element in $H^*(CP(\xi);Z)$ can be written in the form $\pi^*(x_0) + a\pi^*(x_1) + \ldots + a^{n-1}\pi^*(x_{n-1})$ for a unique choice of elements $x_0,\ldots,x_{n-1}$ in $H^*(X;Z)$.

There is the bundle of vectors parallel to the fibres of $CP(\xi) \to X$. If we write the total Chern class of ξ in factored form $(1+b_1)\ldots(1+b_n)$ then the Chern class of this U(n-1)-bundle over $CP(\xi)$ is

$$(49.1) \quad (1+\pi^*(b_1)-a)\ \ldots\ (1+\pi^*(b_n)-a).$$

Remembering that this is only an (n-1)-plane bundle, however, we see there is a relation

$$(49.2) \quad (a-\pi^*(b_1))\ \ldots\ (a-\pi^*(b_n)) = 0.$$

There is a specific class of U(n)-bundles for which we must know the total Chern class. Let (S^1,X) denote a principal $S^1 = U(1)$ action. Then on the quotient there is a corresponding Chern class $a \in H^2(X/S^1;Z)$ On $X \times C^k$ we introduce an action of S^1 by $t(x,z_1,\ldots,z_k) = (tx,t^{-n_1}z_1,\ldots,t^{-n_k}z_k)$. This defines a complex k-plane bundle $(X \times C^k/S^1 = \xi \to X/S^1$.

(49.3) LEMMA: *The total Chern class of* $\xi \to X/S^1$ *is given by* $(1+n_1a)\ \ldots\ (1+n_ka)$.

PROOF: There is the line bundle with Chern class a, $\eta \to X/S^1$ given by $t(x,z_1) = (tx,t^{-1}z_1)$ and $(X \times C)/S^1 \to X/S^1$. Plainly $\xi \to X/S^1$ is $\eta^{n_1} \oplus \eta^{n_2} \oplus \ldots \oplus \eta^{n_k}$. ∎

Now we can proceed to some examples. For S^1 actions we plan to set up a construction analogous to that of Section 25 which was used to introduce the operator Γ. We shall follow two lines of approach.

Consider a smooth action (τ, M^n) of S^1 on a closed oriented manifold. We think of $D^2 \subset C$ as all complex numbers with $z\bar{z} \leqq 1$. Then on $D^2 \times M^n$ we define two actions τ_1 and τ_2 by

$$t(z,x) = (tz,x) \quad \text{for} \quad \tau_1$$

$$t(z,x) = (tz,tx) \quad \text{for} \quad \tau_2.$$

Restricting to $(\tau_1, S^1 \times M^n)$ and $(\tau_2, S^1 \times M^n)$ we obtain two actions which are equivariantly diffeomorphic. Indeed the equivariant diffeomorphism is $\varphi(t,x) = (t,tx)$. Now from the disjoint union $(\tau_1, D^2 \times M^n) \sqcup (\tau_2, -D^2 \times M^n)$ we form a closed oriented $(n+2)$-manifold M^{n+2} together with an action of S^1, (τ, M^{n+2}). We just use φ along the boundaries to glue together.

Thus given (τ, M^n) we canonically construct (τ, M^{n+2}). In $(\tau_1, D^2 \times M^n)$ the S^1 acts freely on the compliment of $\{o\} \times M^n$ and leaves every point of $\{o\} \times M^n$ fixed. In $(\tau_2, D^2 \times M^n)$ the S^1 acts freely on the complement of $\{o\} \times M^n$ and at every point (o,x) the isotropy subgroup is just what it was at x in (τ, M^n). Thus the singularities of the action (τ, M^{n+2}) as well as their normal bundles are quickly catalogued from knowing those of (τ, M^n). For example, the reader may show

(49.4) EXERCISE: *Consider a smooth action* (τ, M^n) *of* S^1 *on a closed oriented manifold and the associated action* (τ, M^{n+2}). *There are maps* $T : M^n \to M^n$ *and* $T' : M^{n+2} \to M^{n+2}$ *of period* p *given by* $T(x) = \lambda x, T'(x) = \lambda x$ *where* $\lambda = \exp(2\pi i/p)$. *If the fixed point set of* (T, M^n) *has trivial normal bundles, then so does the fixed point set of* (T', M^{n+2}). ∎

Note here that the fixed point set of T' always includes a copy of M^n.

Now we shall look at another description of (τ, M^{n+2}). Consider the action of S^1 on $S^3 \times M^n$ given by $t[(z_1,z_2),y] = [(tz_1,tz_2),t^{-1}y]$ and take the orbit space $(S^3 \times M^n)/S^1$. This is a closed oriented $(n+2)$-manifold fibred by M^n over $S^3/S^1 = CP(1)$ with structure group S^1. Next, S^1 will act on $(S^3 \times M^n)/S^1$ by $t[(z_1,z_2),y] = [(tz_1,z_2),y]$. Call this $(\tau', (S^3 \times M^n)/S^1)$.

(49.5) LEMMA: *There is an equivariant diffeomorphism* Θ *of* $(\tau', (S^3 \times M^n)/S^1)$ *onto* (τ, M^{n+1}).

PROOF: Consider a set $A \subset S^3 \times M^n$ given by $[(z_1,z_2),y]$ with $|z_1| \leq |z_2|$. Define $\Theta_1 : A \to D^2 \times M^n$ by $\Theta_1[(z_1,z_2),y] = (z_1/z_2,(z_2/|z_2|)y)$. Then $\Theta_1([tz_1,tz_2],t^{-1}y) = \Theta_1([z_1,z_2],y)$ and thus Θ_1 induces $\hat{\Theta}_1 : A/S^1 \to D^2 \times M^n$. Moreover, $\hat{\Theta}_1 : (\tau', A/S^1) \to (\tau_1, D^2 \times M^n)$ is an equivariant diffeomorphism.

There is also $B \subset S^3 \times M^n$ given by $[(z_1,z_2),y]$ with $|z_1| \geq |z_2|$. Define $\Theta_2 : B \to D^2 \times M^n$ by $\Theta_2([z_1,z_2],y) = (z_1|z_2|^2/|z_1|^2 z_2,(z_1/|z_1|)\cdot y)$. This will induce $\hat{\Theta}_2 : B/S^1 \to D^2 \times M^n$ and $\hat{\Theta}_2 : (\tau', B/S^1) \to (\tau^2, D^2 \times M^n)$ is an equivariant diffeomorphism. Finally, on $A \cap B$ we have $\varphi\Theta_1 = \Theta_2$ where $\varphi : (\tau_1, S^1 \times M^n) \simeq (\tau_2, S^1 \times M^n)$ was described earlier. There results then an equivariant diffeomorphism $\Theta : (\tau', (S^3 \times M^n)/S^1) \simeq (\tau, M^{n+2})$ as required. ∎

We shall again use $\Gamma(\tau,M^n)$ to denote the (τ,M^{n+2}). Iterating this construction we get a sequence of manifolds M^{n+2k} together with actions of S^1. There is an explicit formula

(49.6) $M^{n+2k} = ((S^3)^k \times M^n)/T^k$

where the k-torus acts by

$$(t_1,\ldots,t_k)((z_1,w_1),\ldots,(z_k,w_k),y) =$$

$$((t_1z_1,t_1w_1),(t_1^{-1}t_2z_2,t_2w_2),\ldots(t_{k-1}^{-1}t_kz_k,t_kw_k),t_k^{-1}y).$$

A principal action $(T^k,(S^3)^k)$ is also obtained by just omitting the y. The quotient is $L^{2k} = (S^3)^k/T^k$. The classifying map of this principal T^k-bundle induces

$$\rho : H^*(B(T^k);Z) \to H^*(L^{2k};Z).$$

We consider $H^*(B(T^k);Z)$ as a polynomial ring $Z[a_1,\ldots,a_k]$. We denote $\rho(a_j)$ by $a_j \in H^2(L^{2k};Z)$ also.

(49.7) LEMMA: *The cohomology ring $H^*(L^{2k};Z)$ is generated by the elements $a_1,\ldots,a_k$, all of degree two. These are subject to the relations $a_1^2 = 0$ and $a_j^2 = a_ja_{j-1}$, $2 \leq j \leq k$. In particular $H^{2k}(L^{2k};Z)$ is generated by $(a_k)^k = a_1a_2\ldots a_k$.*

PROOF: For $k = 1$, $L^2 = CP(1)$ and the assertion is obvious. We can proceed inductively. Assume the result is valid for $k - 1$. We let $T^k = T^{k-1} \times S^1$ then the projection $(S^3)^k \to (S^3)^{k-1}$ given by $((z_1,w_1),\ldots,(z_k,w_k)) \to ((z_1,w_1),\ldots,(z_{k-1},w_{k-1}))$ is equivariant with respect to the T^{k-1}-actions. This will induce

$$(S^3)^k/T^{k-1} \to (S^3)^{k-1}/T^{k-1} = L^{2k-2}$$

which is a fibre bundle with fibre the 3-sphere. Actually this is the sphere bundle associated with the complex 2-plane bundle

$$\xi = ((S^3)^{k-1} \times C^2)/T^{k-1} \to L^{2k-2}$$

where T^{k-1} acts on C^2 by

$$(t_1,\ldots,t_{k-1})(z,w) = (t_{k-1}^{-1}z,w).$$

But L^{2k} is just $CP(\xi) \to L^{2k-2}$, the associated projective space bundle with fibre CP(1). According to (49.3) the Chern class of $\xi \to L^{2k-2}$ will be $1 + a_{k-1}$. There is also the unique class $a \in H^2(CP(\xi);Z) = H^2(L^{2k};Z)$ for which every element in $H^*(L^{2k};Z)$ can be uniquely expressed as $\pi^*(x_0) + a\pi^*(x_1)$. From (49.2) we then learn that $(a-\pi^*(a_{k-1}))a = o$ so $a^2 = a\pi^*(a_{k-1})$. We put $a = a_k$ and (49.7) follows with $a_j = \pi^*(a_j)$ for $1 \leq j \leq k - 1$.

(49.8) THEOREM: *Consider the action* $(\tau,CP(p-1))$ *of* S^1 *given by* $t[z_1,\ldots,z_p] = [z_1,tz_2,\ldots,t^{p-1}z_{p-1}]$. *By application of the* Γ *construction there results a sequence of closed oriented manifolds* $M^{2(p+k)-2}$ *for* $k \geq o$. *For each of these manifolds all Pontrjagin numbers are divisible by* p. *Those* M^n *with* $n = 2p^j - 2$ *are Milnor base elements for* $MSO_*/pMSO_*$.

PROOF: Cutting back to the map of period p induced by $(\tau,CP(p-1))$ we see that the fixed point set has trivial normal bundles, and so by (49.4) this is also true for the fixed point set of the map of period p on each construction $\Gamma^k(\tau,CP(p-1))$. According to (48.1) each $[M^{2(p+k)-2}]$ belongs to I(p).

Obviously the major part of the theorem is the assertion about the occurrence of Milnor manifolds. We think of CP(p-1) in the usual way as S^{2p-1}/S^1. The manifold $M^{2(p+k)-2}$ can be thought of as $(S^3)^k \times S^{2p-1}/T^{k+1}$ where T^{k+1} acts by

$$(t_1,\ldots,t_{k+1})((z_1,w_1),\ldots,(z_k,w_k),(x_1,\ldots,x_p)) =$$

$$(t_1z_1,t_1w_1,\ldots,(t_{k-1}^{-1}t_kz_k,t_kw_k),(t_{k+1}x_1,t_k^{-1}t_{k+1}x_2,\ldots,t_k^{-p+1}t_{k+1}x_p).$$

Thinking of $T^k \subset T^{k+1}$ as having $t_{k+1} = 1$ we shall have a (2p-1)-sphere bundle

$$S(\xi) : ((S^3)^k \times S^{2p-1})/T^k \to L^{2k}$$

and $M^{2(p+k)-2}$ is exactly the associated $CP(\xi)$. Then every element of $H^*(CP(\xi);Z)$ can be expressed as $\pi^*(x_o) + a\pi^*(x_1) + \ldots + a^{p-1}\pi^*(x_{p-1})$. Appealing to (49.3) the Chern class of the complex vector bundle $\xi \to L^{2k}$ is $(1+a_k)(1+2a_k)\ldots(1+(p-1)a_k)$.

The tangent bundle $\tau \to M^{2(p+k)-2}$ splits into $\tau = \tau_1 \oplus \tau_2$ in which τ_1 is the bundle along the fibres and τ_2 is induced by $\pi : CP(\xi) \to L^{2k}$ from the tangent bundle to L^{2k}. The Chern class $c(\tau_1)$ then is

$$(1-a)(1-a+\pi^*(a_k))\ldots(1-a+(p-1)\pi^*(a_k))$$

and

$$a(a-\pi^*(a_k))\ldots(a-(p-1)\pi^*(a_k)) = o.$$

We shall use b to denote $\pi^*(a_k)$. From (49.7), $a^{p-1}b^k$ generates $H^{2(p+k)-2}(M^{2(p+k)-2};Z)$, while for $j < p-1$, $a^j b^{p+k-j-1} = o$ for dimensional reasons.

Assume $n = 2p^j - 2$ and consider M^n. There is the universal polynomial in the Pontrjagin classes $s_{n/4}(p_1,p_2,\ldots)$. For a vector bundle ξ we le $S_{n/4}(\xi)$ denote $S_{n/4}(p_1(\xi),p_2(\xi),\ldots)$. If the base V^n of ξ is a closed oriented manifold then put

$$S_{n/4}[\xi] = \langle s_{n/4}(\xi),\sigma(V^n)\rangle \in Z.$$

There is also the additivity formula of Thom

$$s_{n/4}(\xi_1 \oplus \xi_2) = s_{n/4}(\xi_1) + s_{n/4}(\xi_2).$$

We shall apply this to the tangent bundle of M^n; that is,

$$s_{n/4}[M^n] = s_{n/4}[\tau_1] + s_{n/4}[\tau_2] \in Z.$$

However,

$$\begin{aligned} s_{n/4}[\tau_2] &= \langle s_{n/4}(p_1(\tau_2),\ldots),\sigma(M^n)\rangle \\ &= \langle \pi^*(s_{n/4}(p_1(L^{2k}),\ldots),\sigma(M^n)\rangle \\ &= \langle s_{n/4}(p_1(L^{2k}),\ldots),\pi^*(\sigma(M^n))\rangle \\ &= o. \end{aligned}$$

Now then $s_{n/4}[M^n] = s_{n/4}[\tau_1]$. We can write the total Pontrjagin class of τ_1 as

$$(1+a^2)(1+(a-b)^2)\ldots(1+(a-(p-1)b)^2)$$

so that

$$s_{n/4}(\tau_1) = a^{p^k-1} + (a-b)^{p^k-1} + \ldots (a-(p-1)b)^{p^k-1}.$$

We want to show $s_{n/4}(\tau_1)$ is equal to $pa^{p-1}b^{p^k-p}$ mod $p^2H^n(M^n;Z)$. Since $a^{p-1}b^{p^k-p}$ generates $H^n(M^n;Z)$ that will complete the argument.

Since

$$o = a(a-b)\ldots(a-(p-1)b) = a^p - b^{p-1}a \bmod p$$

we have $a^p = ab^{p-1} + pc$ for some $c \in H^{2p}(M^n;Z)$. Now then

$$\begin{aligned} a^{p^2-1} &= (a^p)^{p-1}a^{p-1} \\ &= (b^{p-1}a + pc)^{p-1}a^{p-1} \\ &= (b^{p^2-2p+1}a^{p-1} + p(p-1)b^{p^2-3p+2}a^{p-2}c)a^{p-1} \bmod p^2 \\ &= b^{p^2-3p+2}a^{p-3}(b^{p-1}a + pc)(b^{p-1}a + p(p-1)c) \\ &= b^{p^2-p}a^{p-1} \bmod p^2. \end{aligned}$$

Continuing in this vein it will follow that

$$a^{p^k-1} = a^{p-1}b^{p^k-p} \bmod p^2.$$

If $r = 1,\ldots,p-1$ then

$$\begin{aligned} (a-rb)^p &= a^p - r^pb^p \bmod p \\ &= b^{p-1}a - rb^p \bmod p \\ &= b^{p-1}(a-rb) \bmod p. \end{aligned}$$

Now from the above computations

$$(a-rb)^{p^k-1} = b^{p^k-p}(a-rb)^{p-1} \bmod p^2.$$

Then
$$\begin{aligned} s_{n/4}(\tau_1) &= b^{p^k-p}(a^{p-1} + \ldots + (a-(p-1)b)^{p-1}) \bmod p^2 \\ &= pa^{p-1}b^{p^k-p} \bmod p^2. \end{aligned}$$

This completes the argument. ∎

(49.9) COROLLARY: *The ideal of elements in* MSO_* *which contain a representative* M^n *admitting a diffeomorphism* $T : M^n \to M^n$ *whose fixed point set has trivial normal bundles is precisely the ideal,* $I(p)$, *of bordism classes all of whose Pontrjagin numbers are divisible by* p.

PROOF: For half the corollary we appeal to (48.1). For the other half we use (49.8), (49.4) and (47.1). ∎

50. Module structure of $\widetilde{MSO}_*(C_p)$

In effect we are going to show that $\widetilde{MSO}_*(C_p)$ has projective dimension 1 as a module over MSO_*. We take for $k \geq 1$ $\alpha_{2k-1} \in \widetilde{MSO}_{2k-1}(C_p)$ to be the $[T,S^{2k-1}]$ with $T(z_1,\ldots,z_k) = (\lambda z_1,\ldots,\lambda z_k)$ so that the α_{2k-1} generate $\widetilde{MSO}_*(C_p)$ as a module and $\Delta_1\alpha_{2k+1} = \alpha_{2k-1}$.

(50.1) LEMMA: *There is a sequence of closed oriented manifolds* M^{4k} *such that for* $k \geq 0$,

$$p\alpha_{2k+1} + \alpha_{2k-3}[M^4] + \alpha_{2k-7}[M^8] + \ldots = 0$$

in $\widetilde{MSO}_{2k+1}(C_p)$.

PROOF: Such a sequence is not unique, but we can give an inductive procedure and point out the choice to be made at each step. Actually we shall define a sequence of maps of period p on closed manifolds (T,M^{2k}).

To obtain (T,M^2) we recall that $p[T,S^1] = 0$, hence there is a (T,M^2) on a closed oriented 2-manifold with p fixed points, each having a closed invariant neighborhood equivariantly diffeomorphic to the map on D^2 given by $z \to \lambda z$. Suppose (T,M^{2k}) has been defined. Consider $(T_1, D^2 \times M^{2k})$ and $(T_2, D^2 \times M^{2k})$ respectively given by $T_1(z,x) = (\lambda z,x)$ and $T_2(z,x) = (\lambda z,Tx)$. By (37.1), $[T_1, \dot{D}^2 \times M^{2k}] = [T_2, \dot{D}^2 \times M^{2k}]$ in $\widetilde{MSO}_{2k+1}(C_p)$. Therefore we can select (but not canonically) a fixed point free (τ, B^{2k+2}) on a compact oriented manifold with

$$(\tau, \dot{B}^{2k+2}) = (T_1, \dot{D}^2 \times M^{2k}) \sqcup (T_2, -\dot{D}^2 \times M^{2k}).$$

Now define (T, M^{2k+2}) as

$$(T_1, D^2 \times M^{2k}) \cup (\tau, -B^{2k+2}) \cup (T_2, -D^2 \times M^{2k})$$

by suitable identifications along the boundaries.

The fixed point set of (T,M^{2k}), with appropriate identifications, is a disjoint union

$$M^{2k-2}, -M^{2k-4}, M^{2k-6},\ldots,(-1)^k M^0.$$

Of course M^0 is p points. Each normal bundle is trivial. Since MSO_n consists only of 2-torsion for n $\not\equiv$ (mod 4) we can write out

$$p\alpha_{2k-1} + \alpha_{2k-5}[M^4] + \alpha_{2k-7}[M^8] + \ldots = 0.$$

This proves the remark. ■

Assume now we have fixed such a sequence M^{4k}. Let C be a free graded MSO_* with generators $\gamma_1, \gamma_2, \ldots$, where $\deg(\gamma_j) = j$. Define an MSO_*-module endomorphism $\partial : C \to C$ with degree -1 by $\partial\gamma_{2k-1} = 0$ and

$$\partial\gamma_{2k} = p\gamma_{2k-1} + [M^4]\gamma_{2k-5} + \ldots .$$

(50.2) LEMMA: *The resulting homology* MSO_**-module,* $H_*(C)$*, is isomorphic to* $MSO_*(C_p)$ *by the correspondence* $\gamma_{2k-1} \to [T, S^{2k-1}] = \alpha_{2k-1}$ *and* $\gamma_{2k} \to 0$.

PROOF: Define submodules $C^{(k)} \subset C$ so that $C^{(k)}$ is generated by $\gamma_1, \ldots, \gamma_k$. Then $\{0\} \subset C^{(1)} \subset \ldots \subset C^{(k)} \subset \ldots \subset C$ is a filtration of (C, ∂). Therefore there is s spectral sequence $\{E^r_{p,q}\}$ for which E is associated with a filtration of $H_*(C)$ and

$$E^1_{p,1} \simeq H_{p+q}(C^{(p)}/C^{(p-1)})$$

We immediately see then that

$$E^2_{2k,q} = 0, \quad E^2_{2k-j,q} \simeq Z/pZ \otimes MSO_q.$$

Just as in the bordism spectral sequence this sequence too will collapse since $E^2_{p,q} \neq 0$ if and only if p is odd and $q \equiv 0$ (mod 4). Not it can be seen that $H_n(C)$ and $\widetilde{M}SO_n(C_p)$ have the same order (refer to Section 36). ■

Now suppose C_0 to be the free module generated by the γ_{2k-1} and C_1 to be the free module generated by the γ_{2k}. We have then a short exact sequence

$$0 \to C_1 \xrightarrow{\partial} C_0 \xrightarrow{\varepsilon} \widetilde{M}SO_*(C_p) \to 0.$$

Here $\varepsilon(\gamma_{2zk-1}) = \alpha_{2k-1} = [T, S^{2k-1}]$ while $\partial\gamma_{2k} = p\gamma_{2k-1} + [M^4]\gamma_{2k-5} + \ldots$. This is a free resolution of $\widetilde{M}SO_*(C_p)$ as an MSO_*-module.

(50.3) THEOREM: *Let* $[M^{4k}]$, $k \geq 1$, *be any sequence of bordism classes which satisfy*

$$p\alpha_{2k-1} + [M^4]\alpha_{2k-5} + \ldots = 0$$

for all $k \geq 1$. *The ideal generated by* p *and all the* $[M^{4k}]$ *is just* I(p), *the ideal of bordism classes all of whose Pontrjagin numbers are divisible by* p.

PROOF: That each M^{4k} belongs to $I_{4k}(p)$ follows from (48.1). The problem is to show that M^{2p^j-2} is a Milnor base element for MSO /pMSO .

Suppose for a moment that for each j fixed we can find closed oriented manifolds $V^0, V^4, \ldots, V^{2p^j-2}$ so that, with V^{2p^j-2} a Milnor manifold,

(50.4) $$\beta_{2p^j-1} = [V^0]\alpha_{2p^j-1} + \ldots [V^{2p^j-2}]\alpha_1$$

lies in $p\widetilde{MSO}_*(C_p)$. According to (50.2) if we look at (C,∂) where ∂ is defined by the sequence $[M^{4k}]$ we must have

$$\partial([W^0]\gamma_{2p^j} + \ldots + [W^{2p^j-2}]\gamma_2) =$$

$$px + [V^0]\gamma_{2p^j-1} + \ldots + [V^{2p^j-2}]\gamma_1$$

for a suitable $x \in C$ and bordism classes $[W^n]$. Hence modulo $pMSO_{2p^j-2}$ we see that $[V^{2p^j-2}] = [W^0][M^{2p^j-2}] + [W^4][M^{2p^j-1}] + \ldots$ Since $[V^{2p^j-2}]$ is a Milnor base element of $MSO_*/pMSO_*$ it will follow that $[M^{2p^j-2}]$ is also.

Now we must show (50.4) does occur. Actually this was done in Section 49. What happened there was this. Consider V^0 = p points, V^{2p-2} = $CP(p-1)$ and the action (S^1, V^{2p-2}). By repeated application of the Γ-construction we arrive at relations

$$[V^{2(p+k)-2}]\alpha_1 - [V^{2(p+k)-4}]\alpha_3 + \ldots + (-1)^k[CP(p-1)]\alpha_{2k-1}$$
$$(-1)^k p[T, S^{2(p+k)-1}].$$

The action $(T, S^{2(p+k)-1})$ is given by

$$T(x_1,\ldots,x_{p-1},z_1,\ldots,z_{k+1}) = (\lambda x_1,\ldots,\lambda^{p-1}x_{p-1},\lambda z_1,\ldots,\lambda z_{k+1}).$$

As we saw, $[V^{2p^j-2}]$ is a Milnor base element for $MSO_*/pMSO_*$. ∎

INDEX OF KEY WORDS AND PHRASES

REFERENCES

Ax J.C. Alexander, On the bordism ring of manifolds with involution, Proc. Amer. Math. Soc. vol. 31 (1972), 536-542.

A M.F. Atiyah, Bordism and cobordism, Proc. Cambridge Philos. Soc. vol. 57 (1971), 200-208.

AB - and R. Bott, A Lefschetz fixed point formula for elliptic complexes, II, applications, Ann. of Math. (2) vol. 88 (1968), 451-491.

ASe_1 - and G. Segal, The index of elliptic operators, II, Ann. of Math. (2) vol. 87 (1968), 531-545.

ASe_2 - and -, Equivariant K-theory and completions, J. Differential Geo. vol. 3 (1969), 1-18.

Av V.G. Averbruch, Algebraic structure of the group of intrinsic homology, Doklady Akad. Nauk SSSR vol. 125 (1959), 11-14.

BM A.L. Blakers and W.S. Massey, On the homotopy groups of a triad, II, Ann. of Math. (2) vol. 55 (1952), 192-201.

Ba J.M. Boardman, Cobordism of involutions revisited, Proceedings of the Second Conference on Compact Transformation Groups, Part I, Lecture Notes in Mathematics vol. 298 (1972), Springer-Verlag, 131-151.

B_1 A. Borel, La cohomologie mod 2 des certains espaces homogènes, Comment. Math. Helv. vol. 27 (1953), 165-197.

B_2 -, Sur la cohomologie des espaces fibrés principaux et des espaces homogènes des groupes de Lie compacts, Ann. of Math. (2) vol. 57 (1953), 115-207.

B_3 -, Sous groupes commutatives et torsion des groupes de Lie compacts connexes, Tohoku Math. J. vol. 13 (1961), 216-240.

B_4 - et al., Seminar on Transformation Groups, Ann. Study 46, Princeton Univ. Press (1960).

BH - and F. Hirzebruch, On characteristic classes of homogeneous spaces, I, Am. J. Math. vol. 80 (1958), 458-538; II, Am. J. Math. vol. 81 (1959), 315-382.

BS - and J.P. Serre, Groupes de Lie et puissances réduites de Steenrod, Am. J. Math. vol. 75 (1953), 409-448.

Bu D.G. Bourgin, On some separation and mapping theorems, Comment. Math. Helv. vol. 29 (1955), 199-214.

Br_1 G.E. Bredon, Sheaf Theory, Series in Higher Mathematics, McGraw-Hill Book Co. (1967).

Br_2 G.E. Bredon, Introduction to Compact Transformation Groups, Pure and Applied Mathematics vol. 46, Academic Press (1972).

BP E.H. Brown and F.P. Peterson, Relations among characteristic classes, I, Topology vol. 3 (1964) suppl. 1, 39-52; II, Ann. of Math. (2) vol. 181 (1965), 356-363.

C_1 H. Cartan et al., Cartan Seminar Notes, 1950-51, Paris.

C_2 -, Cartan Seminar Notes, 1954-55, Paris.

CF_1 P.E. Conner and E.E. Floyd, Fixed point free involutions and equivariant maps, Bull. Amer. Math. Soc. vol. 64 (1960), 416-441.

CF_2 - and -, Differentiable Periodic Maps, Ergebnisse Series vol. 33, Springer-Verlag (1964).

CF_3 - and -, Maps of odd period, Ann. of Math. (2) vol. 84 (1966), 132-156.

CS - and L. Smith, On the complex bordism of finite complexes, Inst. Hautes Études Sci. Publ. Math. No. 37 (1969), 117-221.

D_1 T. tom Dieck, Bordism of G-manifolds and integrality theorems, Topology vol. 9 (1970), 345-358.

D_2 -, Actions of finite abelian p-groups without stationary points, Topology vol. 9 (1970), 359-366.

D_3 -, Lokaliezierung aequivarianter kohomologie Theorien, Math. Z. vol. 121 (1971), 253-262.

D_4 -, Characteristic numbers of G-manifolds I, Invent. Math. vol. 13 (1971), 213-224; II, J. Pure and Applied Algebra, vol. 4 (1974), 31-39.

D_5 -, Periodische Abbildungen unitärer Mannigfaltigkeiten, Math. Z. vol. 126 (1972), 275-295.

Do_1 A. Dold, Demonstration elementaire de deux resultats du cobordism, Ehresmann Seminar Notes 1958-59, Paris.

Do_2 -, Erzeugende der Thomschen Algebra N, Math. Z. vol. 65 (1956), 25-35.

Do_3 -, Vollständigkeiten der Wuschen Relationen zwischen Stiefel-Whitneyschen Zahlen differenzierbarer Mannigfaltigkeiten, Math. Z. vol. 65 (1956), 200-206.

E S. Eilenberg, On the problems of topology, Ann. of Math. (2) vol. 50 (1949), 247-260.

ES - and N.E. Steenrod, Foundations of Algebraic Topology, Princeton Univ. Press (1952).

F E.E. Floyd, Actions of $(Z_p)^k$ without stationary points, Topology vol. 10 (1971), 327-336.

H F. Hirzebruch, Neue Topologische Methoden in der Algebraischen Geometrie, Ergebnisse Series vol. 9, Springer-Verlag (1956).

KSt C. Kosniowski and R.E. Stong, Involutions and characteristic numbers, Topology (to appear).

L P.S. Landweber, On the complex bordism of classifying spaces, Proc. Amer. Math. Soc. vol. 27 (1971), 175-179.

M_1 J.W. Milnor, Groups which act on S^n without fixed points, Am. J. Math. vol. 79 (1957), 623-630.

M_2 -, The Steenrod algebra and its dual, Ann. of Math. (2) vol. 67 (1958), 150-171.

M_3 -, On the cobordism ring Ω^*, Notices Amer. Math. Soc. vol. 5 (1958), 457.

M_4 -, Some consequences of a theorem of Bott, Ann. of Math. (2) vol. 68 (1958), 444-449.

M_5 -, On the cobordism ring Ω^* and a complex analogue, Part I, Am. J. Math. vol. 82 (1960), 505-521.

M_6 -, On the relation between differentiable manifolds and combinatorial manifolds, mimeographed notes, Princeton (1956).

M_7 -, Differential Topology, mimeographed notes, Princeton (1958).

M_8 -, Lectures on Characteristic Classes, mimeographed notes, Princeton (1958).

M_9 -, Differential structures on homotopy spheres, mimeographed notes, Princeton (1959).

MZ D. Montgomery and L. Zippin, Topological Transformation Groups, Interscience Publishing Co. (1955).

Mr A.P. Morse, The behavior of a function on its critical set, Ann. of Math. (2) vol. 40 (1939), 62-70.

Mo G.D. Mostow, Equivariant embeddings in Euclidean space, Ann. of Math. (2) vol. 65 (1957), 432-446.

Mu H.J. Munkholm, A Borsuk-Ulam theorem for maps from a sphere to a compact topological manifold, Illinois J. Math. vol. 13 (1969), 167-185.

P R.S. Palais, Imbedding of compact differentiable transformation groups into orthogonal representations, J. Math. and Mech. vol. 6 (1957), 673-678.

Po L.S. Pontrjagin, Characteristic cycles on differentiable manifolds, Math. Sbornik vol. 21 (1947), 233-284.

Q_1 D.G. Quillen, On the formal group laws of unoriented and complex bordism theory, Bull. Amer. Math. Soc. vol. 75 (1969), 1293-1298.

Q_2 -, Elementary proof of some results of cobordism theory using Steenrod operations, Advances in Math. vol. 7 (1971), 29-56.

R V.A. Rohlin, Intrinsic homologies, Doklady Akad. Nauk SSSR vol 89 (1953), 789-792.

Ro_1 D.C. Royster, Aspherical generators of unoriented cobordism, Proc. Amer. Math. Soc. vol. 66 (1977), 131-137.

Ro_2 -, Thesis, LSU, Baton Rouge (1978).

Se G. Segal, Equivariant K-theory, Inst. des Hautes Études Sci. Publ. Math. No. 34 (1968), 129-151.

Sr J.P. Serre, Groupes d'homotopie et classes de groupes abeliens, Ann. of Math. (2) vol. 58 (1953), 258-294.

Sp E.H. Spanier, Infinite symmetric products, function spaces and duality, Ann. of Math. vol. 69 (1959), 142-198.

SpW - and J.H.C. Whitehead, Duality in homotopy theory, Mathematika vol. 2 (1955), 56-80.

S N.E. Steenrod, Topology of Fibre-Bundles. Princeton Univ. Press (1951).

St_1 R.E. Stong, Involutions fixing projective spaces, Michigan Math. J vol. 13 (1966), 445-457.

St_2 -, Notes on Cobordism Theory, Mathematical Notes, Princeton University Press (1968).

St_3 -, Stationary point free group actions, Proc. Amer. Math. Soc. vol. 18 (1967), 1089-1092.

St_4 -, Equivariant bordism and $(Z_2)^k$ actions, Duke Math. J. vol. 37 (1970), 779-785.

St_5 -, Unoriented bordism and actions of finite groups, Mem. Amer. Math. Soc. No. 103 (1970).

St_6 -, A cobordism, Proc. Amer. Math. Soc. vol. 36 (1972), 584-586.

T_1 R. Thom, Espaces fibres en spheres et carrés de Steenrod, Ann. Sci. Ecole Norm. Sup. vol. 69 (1952), 109-182.

T_2 -, Quelques propriétés globales des variétés differentiables, Comment. Math. Helv. vol. 28 (1954), 18-88.

T_3 -, Travaux de Milnor sur la cobordisme, Bourbaki Seminar Notes 1958-59, Paris.

Wa C.T.C. Wall, Determination of the cobordism ring, Ann. of Math (2) vol. 72 (1960), 292-311.

Wh G.W. Whitehead, Generalized homology theories, Trans. Amer. Math. Soc. vol. 102 (1962), 218-245.

W J.H.C. Whitehead, Combinatorial homotopy, I, Bull. Amer. Math. Soc. vol. 55 (1949), 213-245.

Wy H. Whitney, Differentiable manifolds, Ann. of Math. (2) vol. 37 (1936), 643-680.

Wu Wen-Tsun Wu, Les i-carrés dans une variété grassmannienne, C.R. Acad. Sci. Paris vol. 230 (1950), 918-920.

Y_1 C.T. Yang, Continuous functions from spheres to Euclidean spaces, Ann. of Math., vol. 62 (1955), 284-292.

Y_2 -, On the theorems of Borsuk-Ulam, Kakutani-Yamabe-Yujobo and Dyson, I, Ann. of Math. vol. 60 (1954), 262-282.